TUNDISH TECHNOLOGY FOR CLEAN STEEL PRODUCTION

TUNDISH TECHNOLOGY FOR CLEAN STEEL PRODUCTION

Yogeshwar Sahai

The Ohio State University, USA

Toshihiko Emi

Institute of Research of Iron & Steel,
Jiangsu/Sha-Steel, China

World Scientific

NEW JERSEY · LONDON · SINGAPORE · BEIJING · SHANGHAI · HONG KONG · TAIPEI · CHENNAI

Published by

World Scientific Publishing Co. Pte. Ltd.

5 Toh Tuck Link, Singapore 596224

USA office: 27 Warren Street, Suite 401-402, Hackensack, NJ 07601

UK office: 57 Shelton Street, Covent Garden, London WC2H 9HE

British Library Cataloguing-in-Publication Data
A catalogue record for this book is available from the British Library.

ISBN-13 978-981-270-621-8
ISBN-10 981-270-621-6

Printed in Singapore by B & JO Enterprise

Contents

Preface

Steel production of world has increased significantly in the last decade, and has reached over 1.23 billion tons in 2006 from 850 million tons in 1996. Continuous casting of steel is a widely used process and is an important step in steel production. About 90% of the raw steel produced worldwide is continuously cast now. In some parts of the world including the Western Europe, Japan, and USA, continuous casting represents over 97% of steel production. Concurrent with this increase in production levels are stringent quality requirements which have become crucial in the face of progressively increasing machine throughputs and larger product dimensions. As a result, steel cleanliness and strict composition control, together with surface and internal quality of strands, are now becoming the primary concern of steelmakers.

Tundish is the last metallurgical vessel through which molten metal flows before solidifying in the continuous casting mold. During the transfer of metal through the tundish, molten steel interacts with refractories, slag, and atmosphere. Thus, proper design and operation of a tundish are important for delivering steel of correct composition, quality, and temperature. The past three decades or so have seen major advances in tundish technology for clean steel casting. There exist a large volume of important technical papers published in this area, but there was no such book which covered all aspects, from fundamental principles to operational details, of the tundish technology. This book is an attempt in presenting a detailed discussion of tundish technology for clean steel production.

First half of the book deals with the fundamental aspects and theory necessary for understanding the basic concepts of tundish operations. The remainder of the book deals with the operational aspects of the tundish. One chapter is also devoted to recent, emerging, and novel tundish technologies. Thus, the book is sufficiently fundamental to serve as a text book for a graduate course in Process Metallurgy or as an important reference for a metallurgical researcher; at the same time, it is comprehensive enough to contribute to the understanding of scientists and engineers engaged in research, development, or production in steel industry.

The authors are grateful to many colleagues, associates, and students who have contributed in various ways to this book and to the authors' knowledge and understanding of the subject. The authors particularly appreciate and acknowledge contributions of and knowledge gained by late S. Takeuchi, late M. Wolf, R. Boom, Y. Kishimoto, Y. Tozaki, J. Schade, A.K. Sinha, and S. Chakraborty.

YS
TE

Acknowledgement

Financial support for publication of this book was provided by the Iron and Steel Institute of Japan (ISIJ) and Japan Metal Daily (Tekko Shimbun Corporation).

Chapter 1

Introduction

In modern steelmaking and casting plants, steel is produced either in a basic oxygen furnace (BOF) or in an electric arc furnace (EAF). In a BOF, hot metal and scrap are blown by oxygen gas with a flux addition, such as lime, to remove carbon, phosphorus, sulfur, and silicon. A modern EAF produces steel by remelting and refining steel scrap and other raw materials, and also uses oxygen gas injection and lime addition. Fig. 1.1 [1] schematically shows a modern steelmaking and continuous casting facility. The steel melt with dissolved oxygen thus produced is tapped into a ladle, where it is deoxidized with ferroalloys, Fe-Si, Fe-Si-Mn, and/or metallic aluminum. The deoxidation products, such as silica, manganosilicates, alumina, aluminosilicates, aluminates and/or their composites, are largely removed from the melt by flotation. Whenever necessary, the deoxidized melt is further processed in a ladle furnace (LF) to remove any remaining suspended oxide particles (called non-metallic inclusions, or simply inclusions), to lower the sulfur content, and to adjust the melt's chemistry and temperature. Degassing of steel melt is done in vacuum refining facilities (RH, VAD, or VOD) to decrease hydrogen for crack sensitive grades and/or carbon for ultra low carbon grades to meet customer specifications.

The melt is then transferred from the ladle via a tundish into the mold of a continuous casting machine as shown by Yoshida *et al.* in Fig. 1.2 [2], and is solidified as slabs, blooms, or billets. In the last three decades, continuous casting has become a mature technology for the solidification of steel. Today, continuous casting has almost completely replaced ingot casting except for large castings. Continuous casting offers many advantages including better premium cast-metal yield, chemical homogeneity, and better inclusion cleanliness. In continuous casting process, the tundish plays an important role in linking the ladle with the continuous casting machine. A vast amount of published literature exists on the technology of continuous casting with a tundish, but a comprehensive description of tundish technology from both a fundamental and practical point of view is still lacking.

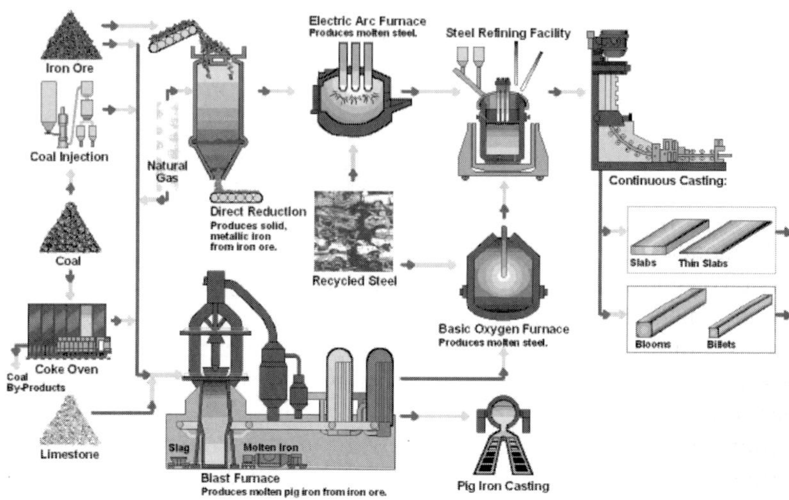

Figure 1.1: A typical steelmaking and continuous casting facility. [Ref. 1]

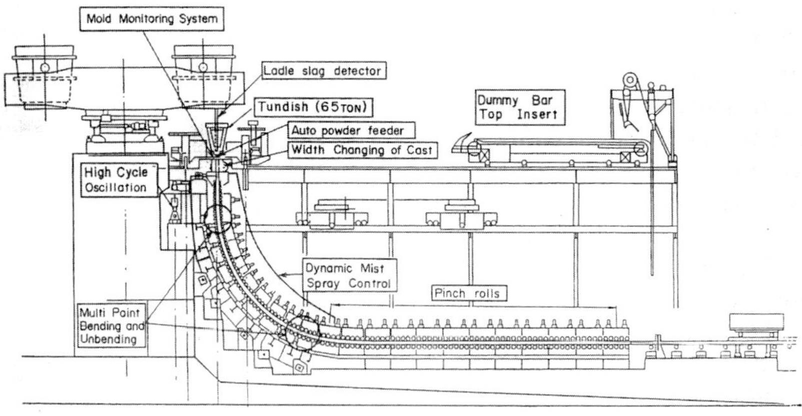

Figure 1.2: Continuous casting machine layout with tundish being installed between the ladle and mold. [Ref. 2]

The aim of this book is to give an overview of tundish technology as an important component of the steel production processes, with emphasis placed on the metallurgical aspects of producing clean steel. The first half of the book presents the fundamental and theoretical aspects of understanding tundish technology. The remainder of the book deals with operational aspects of the tundish. One chapter is also devoted to recent, emerging, and novel tundish technologies. Thus, the book is sufficiently fundamental to serve as a textbook for a graduate course on process metallurgy or as an important reference for a metallurgical researcher or plant engineer in a melting and casting plant. The book does not include any discussion of ladle refining and continuous casting, and coverage of tundish hardware details is rather limited.

Chapter 1 briefly reviews the importance of the tundish in transferring clean steel melt into a continuous casting mold; Chapter 2 deals with the thermodynamics and kinetics of the formation and removal of non-metallic inclusions in a tundish; Chapter 3 reviews the fluid flow and turbulence of steel melt in a tundish as influential factors in reducing inclusions; Chapter 4 deals with the fluid flow characterization of steel melt in a tundish; Chapter 5 describes physical and mathematical modeling of the melt flow in a tundish; Chapter 6 gives details of tundish operation; Chapter 7 discusses active melt temperature control in a tundish; and Chapter 8 touches on innovative new tundish technologies.

1.1 Ingot and Continuous Casting of Steel

Continuous casting has gradually replaced ingot casting over the years, reaching 50% of the annual crude steel production in Japan by 1978, in Italy and former West Germany by 1980, in Korea by 1982, in the UK by 1984, and in the USA by 1986. Large ingots for forgings and small lot production of diverse grades of steel are still produced by ingot casting. Great effort has been devoted to improving the surface and internal quality of continuously cast products to obtain a higher premium cast-metal yield. At the same time, continuous casting productivity was increased substantially to keep pace with the increasing raw steel production capacity. Currently, well over 95% of carbon steels and specialty steels are produced through continuous casting.

In ingot casting, a hollow cast iron mold with a square, rectangular, polygonal, or round cross section is set on the cast iron stool. Finished steel melt is poured from a ladle into the mold in two ways. One is from the mold top, called the top pouring into one mold at a time, and the other is from the mold bottom, called the bottom pouring or uphill teeming into single or multiple molds via the spout and runner bricks.

Typical installation of top and bottom pouring is shown by Eisenkolb and Gerling in Fig. 1.3 [3]. The melt stream in top pouring has more exposure to air and hence suffers from reoxidation. As the pouring stream impinges on the melt surface in the ingot mold, it carries reoxidation products and scum back into the bulk as macro inclusions. During mold filling, metal splash adheres to the mold walls and produces surface defects on the ingot skin, which later requires surface conditioning. In bottom pouring, the melt stream exposure to air, the entrainment of scum, and the occurrence of splash are reduced, but the melt stream contact with the refractory in the pouring spout and runner bricks is longer, which results in contamination of steel melt by inclusions of refractory origin. For large ingots for high end use where quality is important, inert gas shrouding or evacuation is employed during top pouring. For high end use small ingots, bottom pouring is common.

In ingot casting, fully Al-deoxidized steel melt is usually cast into a big-end-up mold with a hot top. Hot topping involves the combination of a thermally insulating board around the top periphery of the ingot and the addition of exothermic and thermally insulating powder on the molten metal. Hot topping retards the solidification of the ingot top, and supplies steel melt from the top to the core part of the ingot to fill the shrinkage cavity caused by steel solidification. During ingot solidification, characteristic crystals and segregation occur in different parts of the ingot as shown by Takenouchi in Fig. 1.4 [4].

Equiaxed crystals, being heavier than liquid metal, settle to the bottom of the ingot. They often trap the rising macro inclusions and carry them along. As solidification of the melt proceeds from the wall to the center of the mold, the formation of fine chilled crystals on the peripheral surface of the ingot is followed by the growth of columnar dendrites, which develop into branched columnar dendrites.

Driven by the difference in specific density, solute enriched melt begins to ascend in front of the branched dendrites, leaving inverse V-segregates along the contour of the solidification front. The core part of the ingot is filled with the sediment-equiaxed crystals from the top. As solidification proceeds, volume contraction in the core part causes the intermittent fall of the equiaxed crystals, and the solute enriched melt part fills the created void. The process results in the formation of

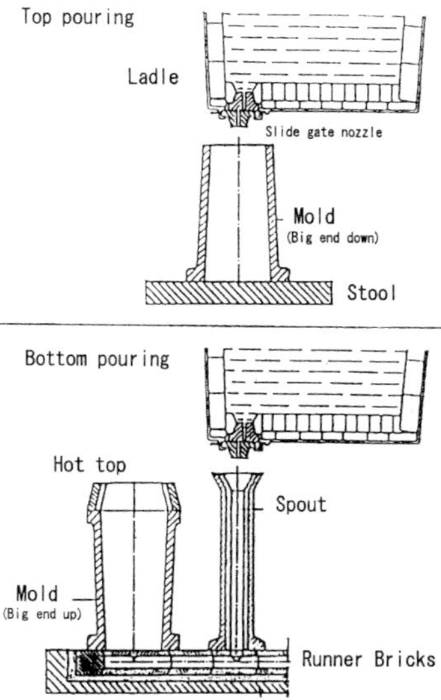

Figure 1.3: Top pouring and bottom pouring for conventional ingot casting. [Ref. 3] [Big-end-up molds for fully deoxidized steels]

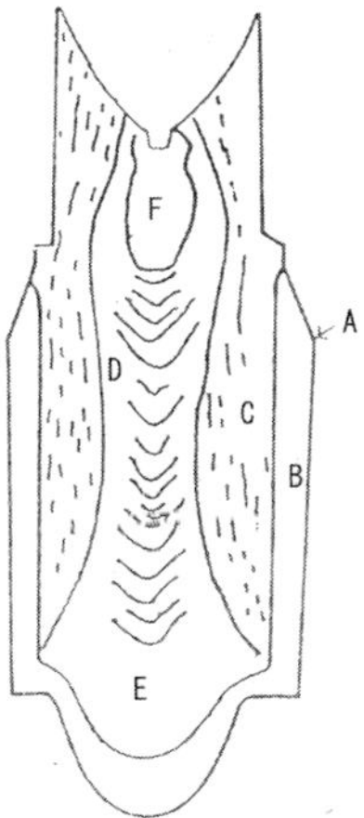

Figure 1.4: Crystal structure and segregation in top poured 75-ton killed steel ingot (A: chilled crystal skin; B: columnar dendrite crystal zone; C: branched columnar dendrite crystal zone with inverse V(or A)-segregates; D: equiaxed dendrite crystal zone with V-segregates; E: sediment equiaxed crystal zone with negative solute segregation and macro inclusions; and F: heavily solute segregated zone). [Ref. 4]

V-segregates, which are often accompanied by porosity or loose structure if the melt supply from the hot top is insufficient or is blocked by the bridging of solidified steel above the core. The degree of solute enrichment and loose structure in the equiaxed dendrites area depends on the superheat of the melt, the ease of melt feeding from the hot top, the height-to-thickness ratio, and the taper of the ingot.

The occurrence of macro inclusions can be reduced by pouring clean steel melt into the mold in a vacuum or under an inert atmosphere. Inverse V- and V-segregates can be minimized by decreasing the solute content (P, S), decreasing Si content to prevent ascend of the interdendritic melt by buoyancy, and optimizing both the mold taper along the longitudinal direction and the aspect ratio of the mold. The occurrence of V-segregates and loose structure can be decreased by increasing the hot topping.

By carefully implementing these measures, Takenouchi of Japan Steel Works has produced 600-ton ingots with virtually no segregates or macro inclusions [4]. These ingots have been used for highly stringent applications, such as the low pressure turbine shaft and pressure vessel for an atomic reactor. In one case, seven heats of decreasing C content were poured in sequence into a 600-ton mold to dilute carbon enrichment in the hot top and in the equiaxed dendrites zone, where V-segregates should appear otherwise. Semi-empirical relationships were developed to decrease the occurrence of inverse V- and V-segregates, and these agreed well with observations. Even in such a careful casting process, however, it became mandatory to crop off the ingot bottom for macro inclusions and the ingot hot top for the solute segregation. This substantially decreased the premium yield of such ingots. Casting ingots under a vacuum in many smaller molds may not be practicable for economic reasons. In addition, the installation and dismantling of molds for ingot casting is a dusty and environmentally unfavorable operation.

Unlike ingot casting, continuous casting gives much better premium cast metal yield, when more than 3 heats are cast sequentially in one campaign. In such a sequence casting, only the very bottom and the very top of the beginning and ending portions, respectively, of each strand are discarded. Compared to normal ingot casting, continuous casting products have much better surface and internal quality, including the segregation and macro inclusions. Inverse V-segregates and V-segregates in the casting direction of continuously cast strands have been reduced to very low levels by active processing, including the following:

(1) Increasing the sedimentation of equiaxed crystals at the pool end of the strand by decreasing casting temperature, by implementing electromagnetic stirring of the melt in the mold to generate nuclei for equiaxed crystal growth, and by dispersing solute enriched melt among the boundaries of the sediment equiaxed crystals;

(2) Preventing suction of the solute rich melt from the surrounding interdendritic area into the pool end in the center of the strands. This is achieved by soft reduction of the strands at the pool end with roll pairs or an anvil pair for an optimized extent (e.g. 0.75 mm/m).

The formation of macro inclusions in continuously cast strands has also been reduced by protecting the pouring stream from air reoxidation during melt transfer from the ladle to tundish by either a long nozzle (a bell type or straight ladle shroud nozzle) or a shrouding pipe. Examples of a long nozzle and shrouded pipe are shown by Shade in Fig. 1.5 [5] and by Yamagami *et al.* in Fig. 1.6 [6], respectively.

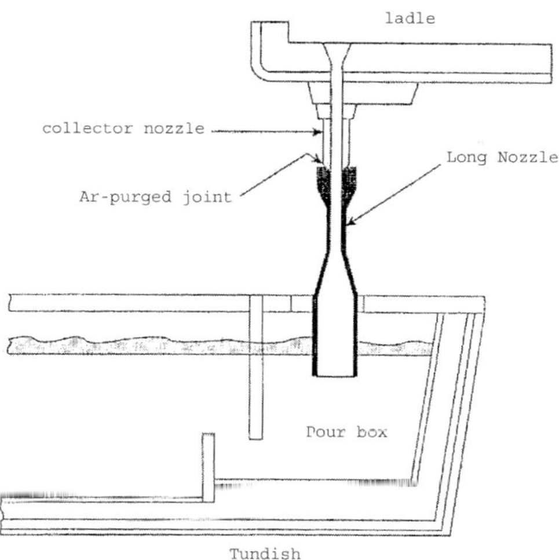

Figure 1.5: Tundish for casting clean steel with a long nozzle for minimum contamination by macro inclusions. [Ref. 5]

The quality of strands has improved, becoming more consistent and more controllable than ingots. Continuously cast strands (semis) are closer in shape to their final products and hence can eliminate the roughing mill in some cases. Further progress of continuous casting technology has made it possible to hot charge as-cast strands into the reheating furnace, or to hot direct roll the as-cast strands. Both techniques avoid surface conditioning, thereby reducing yield loss and earning fuel credit for reheating. Caster productivity and scheduling of casting heats have improved by not requiring surface conditioning of the cast product. Integrated efforts to cast the strands at higher withdrawal rates without compromising the cast metal quality have increased the productivity of some slab casting machines to that of the BOF levels (above 300 kt/mo/machine). In spite of large initial investments, these advantages make continuous casting the preferred solidification process over ingot casting.

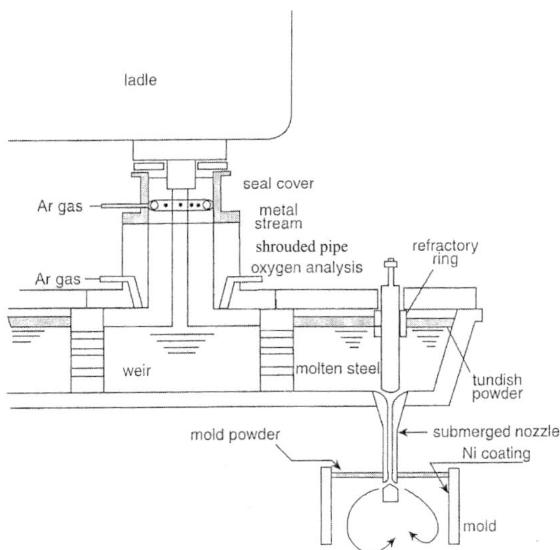

Figure 1.6: Tundish with a shrouded pipe and various devices that are intended to minimize contamination by and maximize flotation of macro inclusions. [Ref. 6]

1.2 The Role of Tundish in the Continuous Casting Process

To transfer finished steel melt from a ladle to the mold in a continuous casting machine, an intermediate vessel, called a tundish, is used. A tundish, as shown by Okimori in Fig. 1.7 [7], is a rectangular big-end-up, refractory-lined vessel, which may have a refractory-lined lid on the top. The tundish bottom has one or more nozzle port(s) with slide gate(s) or stopper rod(s) for controlling the metal flow. The vessel is often divided into two sections: an inlet section, which generally has a pour box and where steel melt is fed from the ladle; and an outlet section from which melt is fed into the mold(s). Various flow control devices, such as dams, weirs, baffles with holes, *etc.*, may be arranged along the length of the tundish. The plan view of different tundish shapes is shown by Wolf in Fig. 1.8 [8]. Dotted lines in Fig. 1.8 indicate melt path from inlet to outlet of the tundish. Longer path is preferred to prolong melt residence time to promote flotation of macro inclusions.

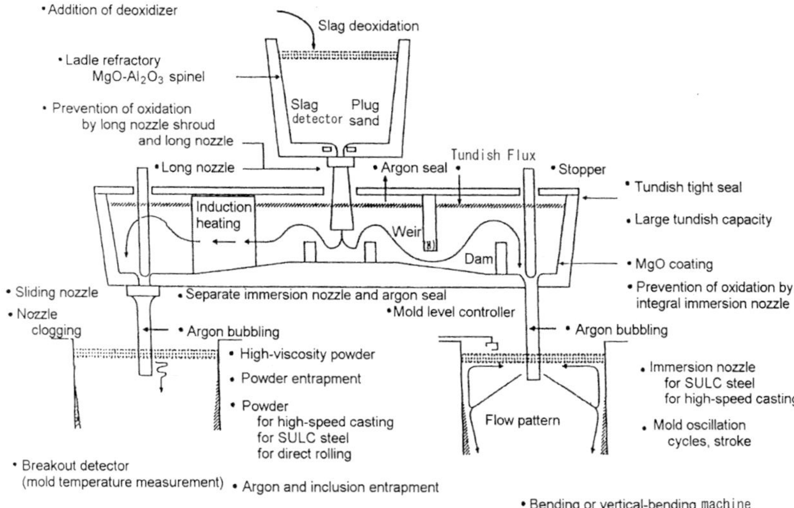

Figure 1.7: Continuous casting system (ladle, tundish, and molds) with all devices and precautions for casting clean steel. (SULC: Super ultra low carbon) [Ref. 7]

The tundish is intended to deliver the molten metal to the molds evenly and at a designed throughput rate and temperature without causing contamination by inclusions. The number of molds is usually 1 or 2 for a slab caster, 2 to 4 for a bloom caster, and 4 to 8 for a billet caster. The melt delivery rate into the mold is held constant by keeping the melt depth in the tundish constant. Any additional delivery rate control is exerted by the slide gates or stopper rods placed at the exit ports of the outlet compartment. The tundish acts as a reservoir during the ladle change periods and continues to supply steel melt to the mold when incoming melt is stopped, making sequential casting by a number of ladles possible. The main causes for inclusion formation and contamination of the melt include reoxidation of the melt by air and carried over oxidizing ladle slag, entrainment of tundish and ladle slag, and emulsification of these slags into the melt. These inclusions should be floated out of the melt during its flow through the tundish before being teemed into the mold.

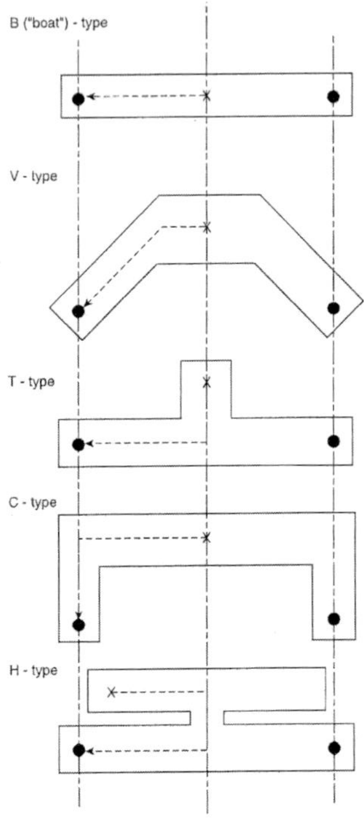

Figure 1.8: Plan view of different tundish shapes. [Ref. 8]

In the past, when ladle metallurgy (ex. ladle furnace, LF) was not fully developed, the tundish was expected to function as a refiner of the deoxidized melt transferred from the ladle where inclusions were not fully removed. Without LF processing, the deoxidized melt had macro inclusions and a large number of micro inclusions of indigenous origin that could agglomerate to form macro inclusions during the melt transfer. A tundish was able to reduce some fraction of macro inclusions from the melt, adjust chemical compositions, and control melt temperature to an appropriate level for feeding into the mold. With the use of the LF and/or degasser, melt cleanliness has significantly improved over the years to meet increasingly stringent customer demands, and the tundish is now seen more as a contaminator than a refiner. Appreciable contamination generally occurred during transient periods (or non steady state) of the sequential casting, i.e., during ladle opening, at the transition of two heats (or ladle change), and during ladle emptying, as shown by Tanaka *et al.* in Fig. 1.9 [9].

During transient periods, the incoming melt stream and any metal splash are heavily reoxidized by the ambient air and by the oxidizing ladle slag that is carried over into the tundish with the melt. The melt stream hits and aggressively emulsifies the ladle slag and tundish slag floating on the melt surface, which eventually get entrained into the melt. Both the reoxidation and the slag entrainment generate harmful macro oxide inclusions. The Al-deoxidized steel melt, even after removal of large particles of deoxidation product in the LF, contains a large number of suspended fine alumina particles. These particles were found to agglomerate by turbulent melt flow during the melt transfer from the ladle via the tundish to the mold, forming large alumina clusters.

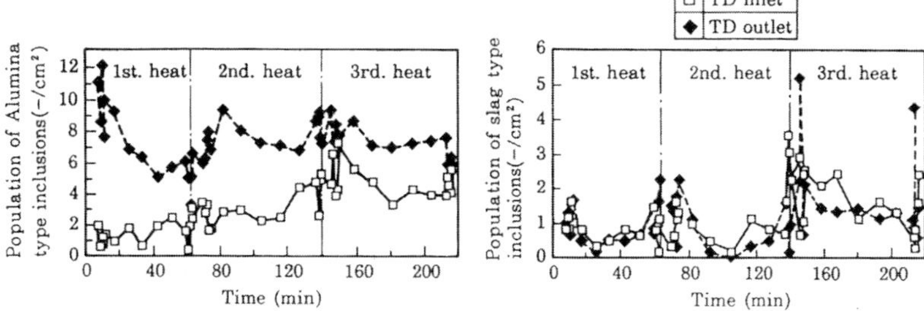

Figure 1.9: Alumina and slag type inclusions at the tundish inlet and outlet (Al-killed steel was poured from a 320-ton ladle with slag containing 5%FeO, into an Ar filled 60-ton tundish with MgO tundish flux). [Ref. 9]

The macro inclusions and large alumina clusters are known to be the major cause of downstream processing problems and defects occurring in strands and their final products. In industry, customers demand cleaner steel with a smaller size of macro inclusions and clusters for better performance of the steel products. Accordingly, the design and operation of a tundish must be directed toward minimizing the formation of the macro inclusions and alumina clusters, and removing them once they form. Otherwise, all the effort made in cleaning the melt in the LF and during other process steps would be of no value.

As shown in Figs. 1.5 through 1.7, various technologies such as a long nozzle or an inert gas shrouding pipe have been implemented to reduce air reoxidation and slag emulsification. Similarly, melt flow control devices have been used to enhance flotation of inclusions formed during the process. Implementation of active control of the melt temperature in a tundish has also contributed to casting clean steel. These measures have proved to be quite successful, at least during the steady state tundish operation, but may not be sufficient for the non-steady state operation, as shown in Fig. 1.9. Non-steady state operation is an integral part of long sequential casting for better metal yield. Although it is desirable to cast steel of high quality, a compromise between the quality and cost has always been struck in any tundish operation.

1.3 The Need for Clean Steel

The requirements for the mechanical properties and chemical composition of steel are constantly increasing, and at the same time the cost, energy, and environmental concerns in steel production are also becoming very important. Thus, the strength, ductility, durability, and corrosion resistance of steel have improved over the years to meet the need. This has been achieved partly by making steel cleaner of non-metallic inclusions, which deteriorate most of the above properties.

Non-metallic inclusions in steel are of two kinds, and each has its different mode of formation. As mentioned earlier, one is indigenous oxide inclusions which form by deoxidation of the steel melt. Most of these oxides are removed during refining and degassing of melt in the ladle, but some non-metallic oxide inclusions of small size remain suspended in the melt. The other kind is exogenous inclusions, which form by reoxidation of deoxidized steel melt by air or by the entrained slag into the melt during the melt transfer from ladle to mold. Usually, inclusions of exogenous origin are much larger than the indigenous ones, and hence are more harmful.

Inclusions cause problems during the casting, rolling, and heat treating processes and sometimes result in failure of the steel during its application. The critical size and composition of the non metallic inclusions that impair the properties of steel are not unique, but depend on the application. Generally speaking, steels with more demanding processing and applications require inclusions smaller in size and number density. Table 1.1, compiled by Emi [10] lists some examples of the critical inclusion sizes and impurity contents for high-end application steel. The critical inclusion size decreases as demands become more stringent.

Another way of reducing the harmful effects of the large inclusions is by modifying the chemical composition of inclusions to lower their melting temperature and to make the inclusions deformable during hot rolling. These large inclusions are elongated in the steel matrix as thin stringers along the rolling direction. Later, when the hot rolled steel is subjected to cold rolling, the thin stringers are broken into pieces of small size because they are brittle at cold rolling temperature. When the distance between the fragmented pieces is made greater by controlling the deformability, an undesirable large inclusion could be split into much smaller, harmless inclusions. This technology is a part of "Inclusion Engineering."

Table 1.1: Critical inclusion sizes and impurity contents tolerable in high performance steels.

Application	Key property	Critical Inclusion Size (μm)	Critical Impurity Content (ppm)
DI-can sheet	Flange crack	<20	
SEDDQ sheet	Average r >2.0		C<20, N<30
Shadow mask	Blur in etching	<5	Low S
Lead frame	Punch crack	<5	
Sour gas pipe	HIC	Shape control	S<5
LNG plate	Embrittlement		P<30, S<10
Lamellar tear	Z-crack	Shape control	*ibid*
Bearing, Race	Rolling-fatigue	<10	O<10, Ti<15
Case hardening	Fatigue crack	<15	O<15, Ti<50
Tire cord	Rupture	Shape control <20	Al<10
Spring wire	Fatigue crack	Shape control <20	*ibid*

Note – DI: Deep drawing & ironing; SEDDQ: Super extra deep drawing quality;
HIC: Hydrogen induced cracking; Z-crack: Crack parallel to rolling direction.

Impurities that dissolve in the melt and form precipitates during solidification need to be minimized as well. Typical examples are phosphorus and sulfur, which form phosphides at the austenite grain boundaries and sulfides in and around the austenite grains. Since it is difficult to remove these impurities in the tundish, they should be minimized during hot metal treatment, the BOF process, and ladle furnace processing before bringing the melt to the continuous casting station.

1.4 Concluding Remarks

An overview of the ingot and continuous casting processes has been presented. The role and functions of the tundish in the continuous casting process and its significance in producing clean steel casting are described:

(1) The tundish links the ladle with the mold of a continuous casting machine. It accepts steel melt from a ladle and delivers it to continuous casting molds with minimum contamination, evenly and at a desired flow rate and temperature;

(2) The tundish is a refractory-lined channel consisting of an inlet and outlet sections and sometimes has flow control devices, such as dams and weirs or a baffle with holes, along its length. A tundish may have a refractory-lined lid, and has bottom ports that are assembled with slide gates or stopper rods through which the melt is teemed into the mold;

(3) Air reoxidation of the incoming steel stream is prevented with the use of a long nozzle immersed into the steel melt in the tundish or by a shrouded pipe with Ar gas flow;

(4) The long nozzle and shrouded pipe also serve to reduce emulsification of the slag into the steel melt; and

(5) Flow control devices in the tundish increase the melt residence time and help in reducing macro inclusions originating from air reoxidation and slag emulsification. At the same time, clusters of agglomerated alumina inclusions are reduced by flotation of these inclusions.

Details on these issues and topics are discussed in the following chapters.

References

1. AISI, Washington, DC, USA, <http://www.steel.org>, accessed January 2006.
2. K. Yoshida, T. Kimura, T. Watanabe, T. Mishima, and M. Ohara, *Tetsu-to-Hagane*, 1988, 74, No. 7, 1240-1247.
3. J. Eisenkolb and R. Gerling, *Proceedings of 7th Ingot Metallurgy Forum* ed. A. A. Tzavarus, May 1994, Pittsburgh, USA, 81-109.
4. T. Takenouchi, *Japan Steel Works Technical Report*, March 1992, No. 66, 1-17.

5. J. Shade, *Lecture Notes, ISS Short Course on Ladle and Tundish Metallurgy for Clean Steels,* Oct. 1997, 314-321.
6. A. Yamagami *et al.*, *Characteristics of Large Cross-Section Bloom Caster for Seamless Tubular Products in the Shrouding of Steel Flow for Casting and Teeming,* Ed. G. Harry ISS, Warrendale, Pa. USA, 1986, 61-71.
7. M. Okimori, *Nippon Steel Technical Report*, 1996, No. 361, 67-76.
8. M. Wolf, *Slab Caster Tundish Configuration and Operation-A Review, Proceedings Steelmaking Conference*, 1996, 79, 367-381.
9. H. Tanaka, R. Nishihara, I. Kitagawa and R. Tsujino, *Tetsu-to-Hagane*, 1993, 79, 1254-1259.
10. T. Emi, *Improving Steelmaking and Steel Properties, Fundamentals of Metallurgy*, Ed. S. Seetharaman, Woodhead Publishing, Cambridge, UK, Inst. of Mater., Minerals & Mining, 2005, 503-554.

Chapter 2

Non-Metallic Inclusions

2.1 Introduction

Steel, as a structural material, has advantageous properties such as, strength, ductility, and durability. The ductility includes deep drawability, cold formability, and low temperature toughness, while durability is against wear, fatigue, hydrogen-induced cracking, and stress corrosion cracking. The ductility and durability are significantly impaired by large-sized, non-metallic inclusions in steel. Steel with such inclusions may be termed 'dirty' steels. Only 'clean' steels, with a smaller number of evenly distributed, small-size inclusions, meet increasing demands for better ductility and durability required for stringent applications.

Non-metallic inclusions include oxides, sulfides, nitrides, carbides, and their compounds or composites. Sulfides, carbides, and nitrides precipitate under normal conditions during cooling of steel below solidus temperatures. Small particles of particular oxide inclusions, sulfides, carbides, and nitrides, have been utilized to control microstructure for improving steel properties. However, most of the large oxide inclusions and some sulfide inclusions form while the steel is in liquid state. If they are not removed from the steel melt before solidification, they can cause defects in the casting products, give rise to processing difficulties and failures, decrease productivity, degrade product properties, and reduce premium yield. This chapter deals with the oxide inclusions as they occur during the melt transfer from ladle via a tundish to a continuous casting mold. The sulfide inclusions can be minimized by ladle refining of the steel melt by desulfurizing prior to the melt transfer to tundish, and hence are not discussed in this chapter.

2.2 Origin of Oxide Inclusions

Oxide inclusions arise from exogenous and indigenous origins. Exogenous inclusions are formed during the melt transfer by (1) reoxidation of deoxidized and refined steel melt when it comes into contact with air and oxidizing slag; and (2) entrainment of the reoxidation product, slag, and refractory. Indigenous inclusions form in the ladle from the reaction between oxygen dissolved in the melt and a deoxidizing element such as Al or Si added to the melt. In a deoxidized and refined steel melt, indigenous inclusions are usually small in size, and hence they are less harmful, provided that they do not agglomerate into large ones during the melt transfer. In contrast, exogenous inclusions often exist in large sizes shortly before the melt is delivered into the mold, have limited opportunity to be removed, and hence are more harmful.

The tundish functions include distribution of the deoxidized steel melt from the ladle to molds evenly, at a designed feed rate and temperature. In addition, a tundish is required to (1) prevent the occurrence of large inclusions of exogenous origin; (2) prevent the formation of large agglomerates of indigenous inclusions, and if the complete prevention is not possible; (3) remove them during the melt transfer through the tundish. To achieve these objectives, the melt transfer must avoid reoxidation by ambient air and ladle slag. Transfer of ladle slag into the tundish, and tundish flux into the mold must be avoided to prevent their emulsification and entrainment into the melt.

2.2.1 *Exogenous inclusions*

2.2.1.1 *Definition of macro inclusions*

Deoxidized melt is commonly refined in the ladle to trim temperature and chemical composition and to remove deoxidation products. Ladle refining usually removes most of the large indigenous inclusions, leaving only a small amount of inclusions of up to 20-50 μm in diameter suspended in the melt. Inclusions greater than about 50 μm, often of exogenous origin, may be called macro inclusions. Inclusions smaller than about 50 μm, mostly of indigenous origin, and small agglomerates of the indigenous inclusions may be called micro inclusions.

The above definitions are not unique, but are used for convenience. Inclusions in the cast strands greater than the critical size impair the properties of steel products as listed in Table 2.1, based on Ref. [1]. Such inclusions can be alternatively defined as macro inclusions. However, the critical size has continuously decreased in the last three decades due to increasingly demanding applications of quality steel products. An example of decreasing critical inclusion size in steel sheet from CC slab for a deep-drawn and ironed two piece can (DI-can) is shown in Fig. 2.1 [2]. For the time being, therefore, inclusions greater than 50 μm in diameter in cast strands are defined as macro inclusions. Macro inclusions up to about 200-500 μm can be occasionally found in cast strands.

Table 2.1: Critical size of macro inclusions in cast semis that cause process upsets and product defects.

Steel Products	Critical Size (μm) in Slabs* or Blooms**
Cold Rolled Sheet	240*
DI-Can	50*
UOE-Pipe	200*
ERW-Pipe	140*
Cold Forgings	100**
Steel Cord	30**
Ball Bearing	15*
Typical Composition	Al_2O_3, $CaO\text{-}(MgO)\text{-}Al_2O_3$, $CaO\text{-}Na_2O\text{-}SiO_2\text{-}Al_2O_3$, $CaO\text{-}MgO\text{-}SiO_2\text{-}Al_2O_3$

2.2.1.2 *Origins of exogenous macro inclusions*

Details of the formation or origin of exogenous inclusions are summarized below:

(1) Reoxidation of deoxidized and refined steel melt at the surface of the teeming stream by air ingress from the joints at the ladle/long nozzle, tundish/slide gate, and slide gate/submerged entry nozzle;

(2) Reoxidation of the melt bath surface in tundish by air ingress;

(3) Reoxidation of the melt bath surface by iron oxide, manganese oxide, and/or silica that are contained either in carried over ladle slag, flux added to tundish, or mixture of the two;

(4) Entrainment of the ladle slag into melt in the tundish by vortexing/draining near the end of emptying of the ladle;

(5) Emulsification and entrainment of tundish slag floating on the melt surface by turbulent impinging teeming stream from the ladle during the opening and changing of the ladle;

(6) Entrainment of tundish slag floating on the melt surface by turbulent surface flow or by vortexing towards the outlet of the tundish, particularly when the bath depth in the tundish is decreased to a low level during the ladle change period;

(7) Reoxidation of the melt at the melt/refractory interface by less stable oxides contained in tundish glaze, the nozzle, and/or the tundish lining;

(8) Entrainment of the fragments of the ladle glaze, ladle nozzle, long nozzle, tundish lining, and/or submerged entry nozzle due to erosion and detachment by turbulent melt flow;

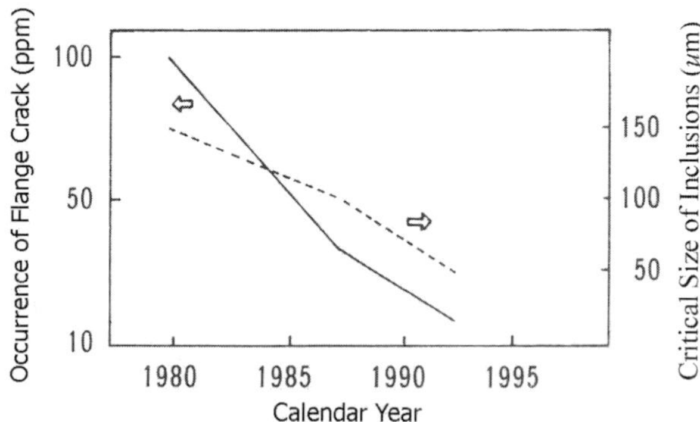

Figure 2.1: Acceptable maximum failure rate of DI cans and related critical size of inclusions in continuously cast slabs. [Ref. 2]

(9) Entrainment of alumina clusters, which are deposited on the inner wall of the tundish nozzle and the SEN and dislodged by turbulent melt flow to the mold.

So it is important to make the ladle melt clean and prevent any ladle slag from flowing into the tundish. Also, careful control of melt teeming from ladle to tundish and from tundish to mold is mandatory to prevent the occurrence of exogenous macro inclusions. This is particularly relevant during the non-steady state casting that include the opening, changing, and emptying of ladles when deviation from the standard operation can happen.

Among the above, origins (1-6), and (9) are more serious since the size of the resulting inclusions can be large, and the inclusions are given limited opportunities for their flotation and removal. Their removal tends to be incomplete since it occurs at the later stage of melt transfer, allowing inclusions to be carried over into the mold. If all the inclusions are not brought to the free surface of melt in the mold, they are entrapped in the solidifying shell.

In the mold, the flux may get entrained in the melt by vortex formation or turbulence generated at the mold flux/steel melt interface. This is particularly so when the velocity of the melt stream exiting from the SEN is designed to be high e.g., at high speed casting and/or at high throughput rate. Practical measures that have proven effective for minimizing the macro inclusions in the tundish are presented in detail in Chapter 6.

2.2.1.3 *Constitution of exogenous inclusions*

The chemistry and structure of the exogenous inclusions are complicated, reflecting compositions of the reoxidized melt, the multi-component ladle slag, the tundish flux, mold flux, and tundish refractory. Quite often, some of the less stable oxides in the inclusions are reduced by the deoxidizing elements in the melt. Also, in many cases the inclusions are agglomerated with indigenous inclusions. Thus, exogenous inclusions exhibit a multi-phase and complex structure. For example, crystalline phases occur embedded in or bound by a glassy phase. Sometimes, one can identify the origin of some of the exogenous inclusions by analyzing their chemical composition and crystallographic structure.

To quantify the amount and identify the origin of macro inclusions, a "slime method" was developed by Yoshida and Funahashi [3]. The slime method extracts macro inclusions by potentiostatic electrolysis of a few to ten kilograms of bomb specimen of the steel melt in a neutral, non-aqua solution, and magnetically separates the macro inclusions in the residue (slime) from iron carbides and steel fragments. The number density of macro inclusions in steel is usually so low that such a large quantity of the specimen is required to make the results statistically meaningful. The macro inclusions are ultrasonically sieved, examined for size distribution by an optical microscope or Coulter counter, and analyzed for chemical composition of the phases in an individual inclusion by electron probe micro analyzer (EPMA). Average chemical composition of each size range of the extracted inclusions is determined by fusing them into a glass-bead, and analyzing the bead by x-ray diffraction spectroscopy (XRF).

Habu *et al.* [4] made an early attempt to quantify the origin of macro inclusions by adding tracers such as La_2O_3 to the tundish flux and Ce_2O_3 to the mold flux under extreme casting conditions, as given in Table 2.2. The CaO, La, and Ce contents in macro inclusions extracted by the slime method were analyzed to determine the contributions of the ladle slag, tundish flux, and mold flux to form the inclusions. The amount of the macro inclusions extracted by Habu *et al.* is presented in Fig. 2.2. Melt reoxidation by air was estimated from reoxidation loss of Al in the melt and the data in Fig. 2.2. In low carbon Al-killed (LCAK) and medium carbon Al-killed (MCAK) HSLA steels cast by a deep tundish with a long nozzle and Ar-shield, the main source of the macro inclusions was found to be air reoxidation seconded by tundish flux entrainment. The ratio of the reoxidation to slag entrainment depended on the tundish operation. Tanaka *et al.* [Ref. 9 in Chapter 1] later reported the factors that influenced the ratio of the number density of macro inclusions ($>10\mu m$ as observed by microscope) of reoxidation origin to that of slag entrainment origin for sequential casting. A part of their study is shown in Fig. 1.9.

Table 2.2: Casting Conditions of 100 ton BOF Melt for LCAK and MCAK Steel via a 35-ton Tundish. [Ref. 4]

Heat No.	Ar gas shroud	Long Nozzle	Tundish	
			Depth	Dams
A	✕	✕	Shallow	✕
B	✕	◯	//	✕
C	◯	◯	//	✕
D	✕	✕	Deep	✕
E	✕	◯	//	✕
F	✕	✕	//	◯
G	✕	◯	//	◯
H	◯	◯	//	◯

(◯······equipped, ✕······not-equipped)

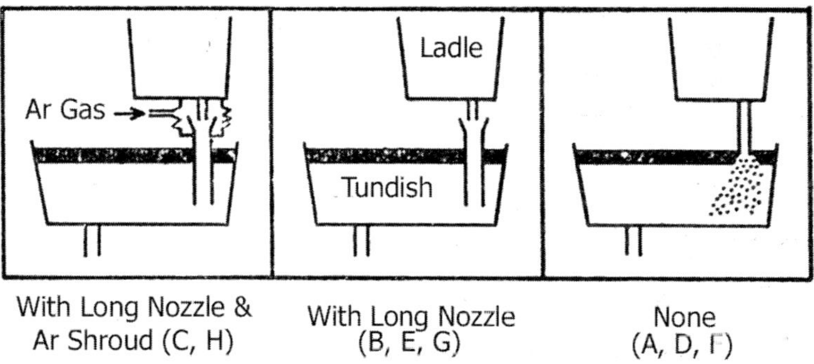

Thus, macro inclusions occurring during the melt transfer contain various ratio of the following constituents: (1) deoxidation- and reoxidation-products, i.e., Al_2O_3 for Al killed steel or manganese aluminosilicate for Si-Mn killed steel; (2) BOF slag of CaO-MgO-SiO_2-Fe_tO system with $(CaO+MgO)/SiO_2 = 3{\sim}8$ that is carried over from ladle into tundish; (3) ladle slag for secondary refining, usually of

CaO-Al$_2$O$_3$-SiO$_2$ system (low in SiO$_2$ for Al-killed steels, low in Al$_2$O$_3$ for Si-killed steels); (4) tundish flux of CaO-Al$_2$O$_3$-SiO$_2$ system with a high ratio of CaO/(Al$_2$O$_3$+SiO$_2$); and (5) mold flux consisting of CaO-SiO$_2$-Al$_2$O$_3$-(NaF) with CaO/SiO$_2$ ≈ 1.

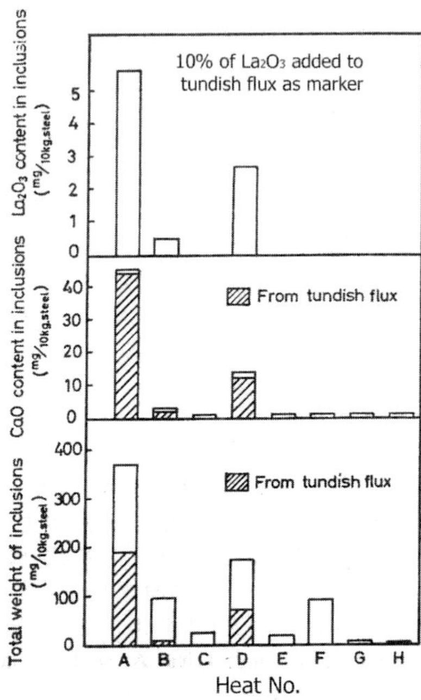

Figure 2.2: Occurrence of macro inclusions of reoxidation and slag entrainment origin under various casting conditions shown in Table 2.2. [Ref. 4]

2.2.2 Indigenous inclusions

2.2.2.1 Thermodynamics of deoxidation reaction

Indigenous inclusions occur as a consequence of the deoxidation reaction

$$n[M] + m[O] = M_nO_m \qquad (2.1)$$

where [M] denotes a deoxidizing element, such as Mn, Si, or Al dissolved in a steel melt. When M is added to the melt as pure metal like Al, or as an alloy like ferromanganese, ferrosilicon, or silico-manganese, significant supersaturation occurs at the periphery of the melting Al or alloys. Nucleation and growth of the resulting oxides, M_nO_m (MnO, SiO_2, MnO-SiO_2, or Al_2O_3), form liquid globular inclusions, irregular solid inclusions, and/or clusters of the solid inclusions. In Eq. (2.1), the dissolved oxygen, [O], in the melt comes from the steelmaking process where oxygen gas is injected into high carbon iron melt to reduce carbon to meet specifications.

The deoxidation reaction in liquid steel is driven by supersaturation, and usually proceeds until chemical equilibrium of Eq. (2.1) is reached to fulfill thermodynamic requirements given by

$$\Delta G^0 = -RT \ln K \tag{2.2}$$

$$K = a_{M_nO_m} / (\gamma_M)^n [M]^n (\gamma_O)^m [O]^m \tag{2.3}$$

Here, ΔG^0 is the Gibbs energy change for Eq. (2.1), R is the gas constant, square brackets indicate concentration in steel melt, and γ is the activity coefficient of the subscript component. Revised values of [O] and [M] calculated by Sakao [5] with recommended values of ΔG^0 for typical deoxidation reactions are shown in Fig. 2.3. Note that revision of ΔG^0 is still in progress, particularly for strong deoxidizing elements as shown partly for Al, Ca, and Mg in the following. Also, the activity values determined by Chipman and coworkers [6, 7] for a strongly and weakly oxidizing component (Fe_tO and SiO_2) in typical BOF slag and ladle slag are given in Figs. 2.4 and 2.5.

Deoxidation with Al is most often employed to produce high quality flat products, and hence equilibrium between [Al] and [O] has been a great concern for steel producers. However, data reported for the deoxidation equilibrium have a large scatter. Recent investigation by Itoh *et al.* [8] recommended that curve 4 in Fig. 2.6 is most reliable. More scatter in the data was observed for the deoxidation equilibria between [Ca] and [O], and [Mg] and [O]. Re-examination of these equilibria was also made by the same authors [8], resulting in Figs. 2.7 and 2.8 where line 3 in each figure is considered reliable.

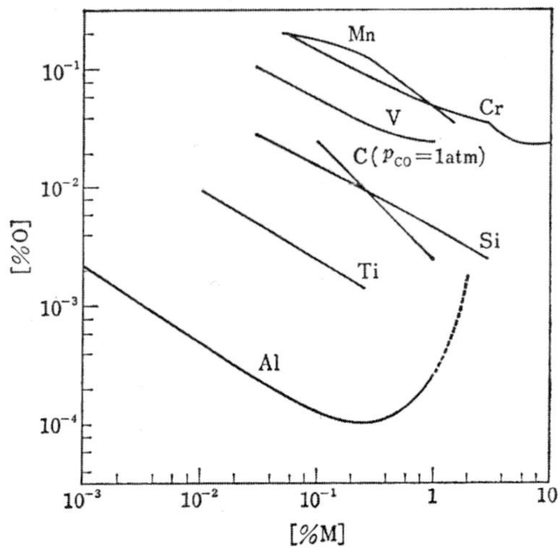

Figure 2.3: Revised deoxidation equilibria with various deoxidation elements in liquid iron at 1873K. [Ref. 5]

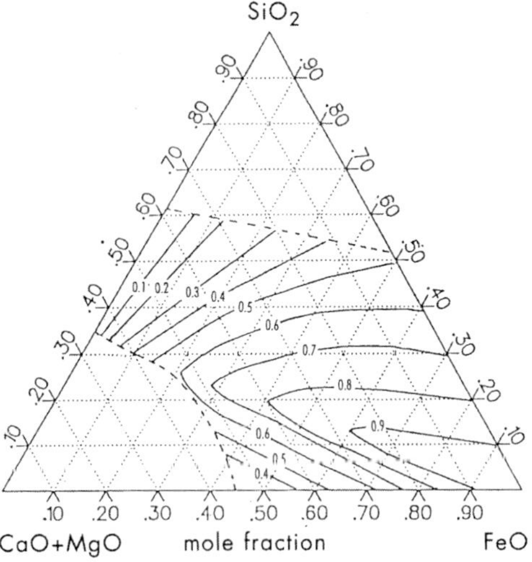

Figure 2.4: Activity of FeO in CaO-FeO-SiO$_2$ slags at 1873K. [Ref. 6]

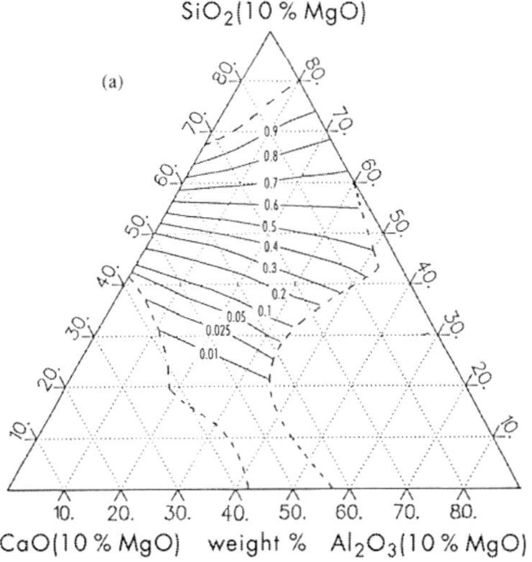

Figure 2.5: Activity of SiO$_2$ in CaO-SiO$_2$-Al$_2$O$_3$-10 mass% MgO slag at 1873K. [Ref. 7]

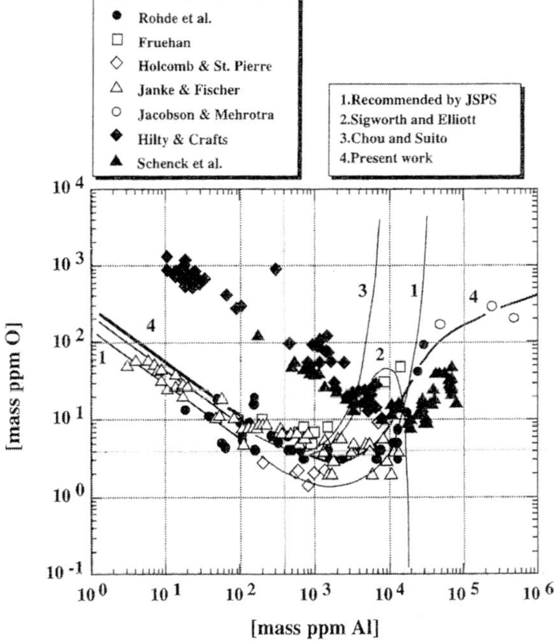

Figure 2.6: Deoxidation equilibria with Al in liquid iron at 1873K. [Ref. 8]

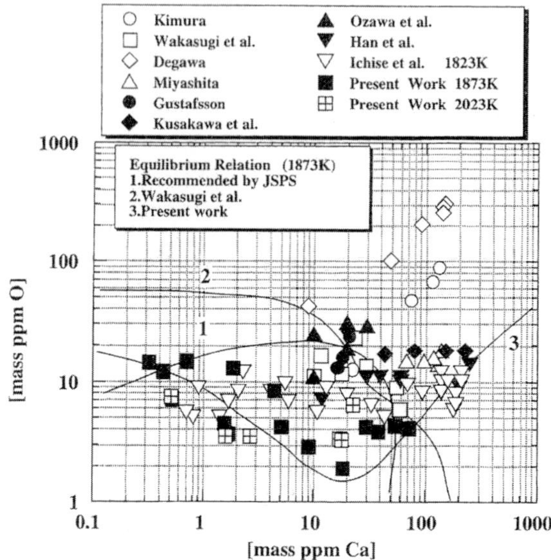

Figure 2.7: Deoxidation equilibria with Ca in liquid iron at 1873K. [Ref. 8]

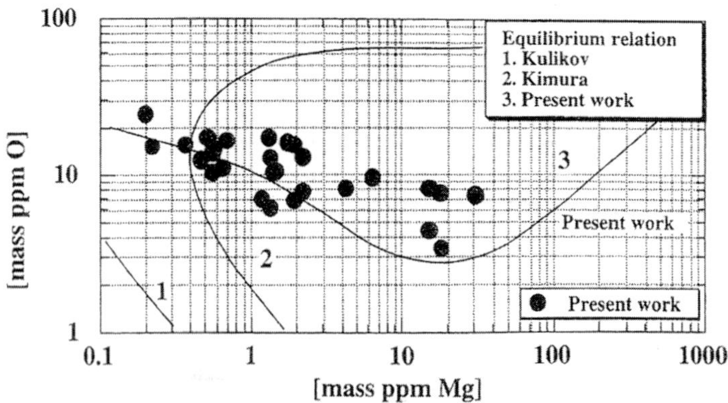

Figure 2.8: Deoxidation equilibria with Mg in liquid iron at 1873K. [Ref. 8]

When Al-deoxidized melt is held in contact with a MgO containing refractory, or Mg-deoxidized melt is held in contact with an Al_2O_3 containing refractory (e.g., the ladle- and tundish-lining and SEN), the reaction between MgO and [Al], or Al_2O_3 and [Mg] takes place according to Eq. (2.4) or Eq. (2.6).

$$MgO(s) + 2[Al] + 3[O] = MgO \cdot Al_2O_3(s) \tag{2.4}$$

$$\Delta G_4^0 = -887,960 + 210.88T(J/mol) \tag{2.5}$$

$$Al_2O_3(s) + [Mg] + [O] = MgO \cdot Al_2O_3(s) \tag{2.6}$$

$$\Delta G_6^0 = -110,720 - 93.51T(J/mol) \tag{2.7}$$

Taking into account the spinel formation reaction,

$$MgO(s) + Al_2O_3(s) = MgO \cdot Al_2O_3(s) \tag{2.8}$$

$$\Delta G_8^0 = -20,740 - 11.57T(J/mol) \tag{2.9}$$

together with Eqs. (2.5) and (2.7), Itoh *et al.* [8] derived phase stability for the oxides involved, resulting in Figs. 2.9 and 2.10.

The octagonal magnesioalumina-spinel, $MgO \cdot Al_2O_3$ (s), and Al_2O_3 are both non-deformable solids at steelmaking temperatures, and form hard clusters by colliding and agglomerating with each other in turbulent melt flow during the melt transfer. The cluster formation can be favorably avoided when two non-deformable inclusions are replaced by either non-clustering MgO or deformable lime-magnesioaluminate, CaO-MgO-Al_2O_3, by a proper choice of the combination of [Al], [Mg], and [Ca] is made as shown in Figs. 2.9 and 2.10.

In reality, liquid steel is covered with molten slag and surrounded by the ladle lining refractory which is not a single kind of refractory since zone lining with different refractories is usually practiced. Accordingly, reactions of liquid steel with these oxide materials should be taken into account. These reactions compete with the deoxidation reactions. Also, activities of the deoxidation products in slag and on the refractory in contact with the steel melt must be considered when the concentration of dissolved oxygen is to be calculated. Some of the activity data for slags typically employed in the ladle furnace and tundish are given, after Rein and Chipman [7] and Hino and Ban-ya [9], in Figs. 2.11 and 2.12. Comparison of Fig. 2.6 to Figs. 2.11 and 2.12 shows that equilibrium dissolved oxygen at 1600°C in steel melt, deoxidized with 0.04%Al under a slag cover of the system CaO-Al_2O_3-SiO_2 or CaO-Al_2O_3-CaF_2 with a_{Al2O3} of 0.25 ($CaO \cong 0.4$ and SiO_2 or $CaF_2 \cong 0.1$ mole fraction), can be a low of 2.5 ppm in contrast to 4 ppm in the melt in equilibrium with pure Al_2O_3 ($a_{Al2O3} = 1$).

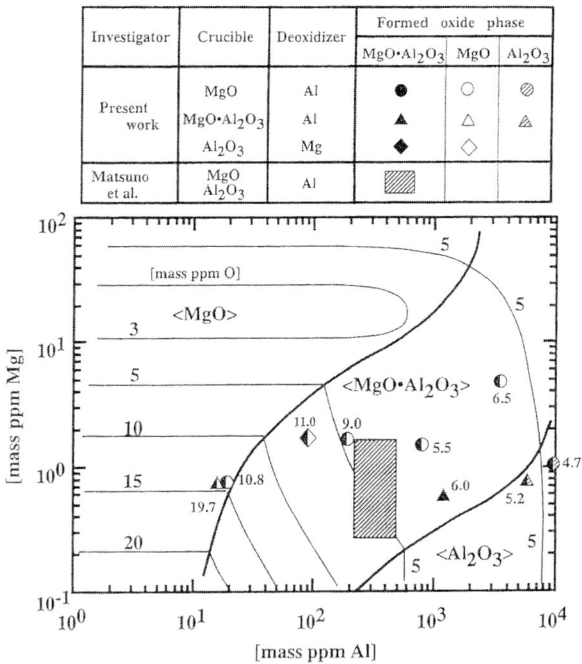

Figure 2.9: Stability of MgO, MgO·Al₂O₃, and Al₂O₃ inclusions, and iso-oxygen contour lines calculated at 1873K with experimental data for steel melt deoxidized with Al and Mg. [Ref. 8]

2.2.2.2 Sequence of oxide inclusion formation

The formation of the above-mentioned oxide inclusions involves nucleation and growth processes. For the nucleation of the inclusions, supersaturation of the solutes involved is usually required. Nucleation can also be of a heterogeneous nature. For a typical example, the actual sequence of the deoxidation reaction with Al and subsequent inclusion formation in the ladle with Al proceeds as follows:

(1) Metallic Al bars are added to the steel melt in the ladle when the melt is teeming from the steelmaking furnace, BOF or EAF. Upon contact with Al bars, the melt solidifies and forms steel shell around the Al bars that are melting;

(2) The shell soon melts away, forming a more or less turbulent and diffuse boundary between the molten Al and the steel melt containing a high concentration of dissolved oxygen;

(3) The chemical reaction between the Al dissolving into the melt and [O] takes place at a high supersaturation of [Al] and [O]. At the same time, dissipation of the supersaturation takes place as the deoxidation reaction and turbulence at the boundary proceed;

(4) Nucleation of Al_2O_3 occurs in the diffuse boundary layer due to the supersaturation. Heterogeneous nucleation also takes place around numerous nuclei suspended in the melt. The nuclei are fine oxide particles;

(5) Growth of the iron oxide-aluminate (hercynite) inclusions occurs only in the very early period of deoxidation when dissolved oxygen content is still high. It is immediately taken over by the growth of alumina on the reduced iron-aluminates or by the nucleation of Al_2O_3. Al_2O_3 inclusions continue to grow by turbulent mass transfer of [Al] and [O] toward the Al_2O_3 nuclei, resulting in large size dendritic, plate-like, and irregular shape Al_2O_3 particles up to a few hundred μm as shown by Dekkers *et al.* [10, 11] in Figs. 2.13 and 2.14;

(6) Enhanced by bulk melt flow and local eddy flow, the nucleation, growth, and simultaneous collision and agglomeration of Al_2O_3 inclusions proceed in the entire melt. Likewise, grown and/or agglomerated large Al_2O_3 inclusions move towards the melt surface by Stokes' flotation which may be assisted by ascending melt flow. The rate of the growth and agglomeration depends on the initial size and number density of Al_2O_3 inclusions, concentrations of [Al] and

[O], and the turbulent dissipation rate of the stirring energy of the melt;

(7) While the number of the large inclusions and the supersaturation decrease with time, the size and shape of Al_2O_3 inclusions that remain or form in the melt, change to smaller spherical and polygonal ones, as shown in Fig. 2.14 (a)-(d). MgO contained in the top slag and refractory is reduced by [Al] to Mg, which reacts with the Al_2O_3 inclusions to form magnesioaluminate, and finally magnesioalumino spinel. The spinel inclusions are subject to collision, agglomeration, and flotation as well, but at much slower rates. The slower rates are due to the dissipation of turbulence of melt flow and also due to the decrease in the number of large Al_2O_3 inclusions that can promote gradient collision with the small-sized spinel inclusions during their Stokes' flotation.

Investigator	Crucible	Deoxidizer	Formed oxide phase			
			$MgO{\cdot}Al_2O_3$	MgO	CaO	$CaO{\cdot}2Al_2O_3$
Present work	Dolomite	Al	■	□	⊠	
Kimura	CaO	Al+Mg		▷	▶	
	MgO	Al+Mg	◀	◁		
	Al_2O_3	Al+Mg	▼			▽

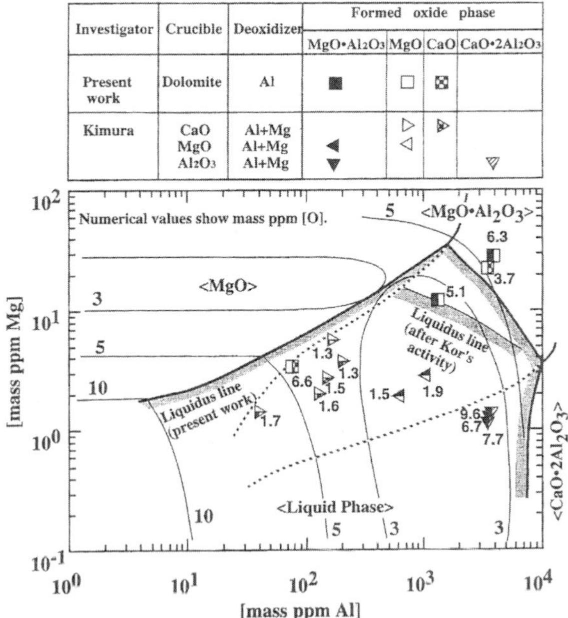

Figure 2.10: Stability of MgO, MgO·Al₂O₃ and CaO·2Al₂O₃, and iso-oxygen contour lines calculated at 1873K with experimental data for steel deoxidized with Al, Mg, and 1 ppm Ca. [Ref. 8]

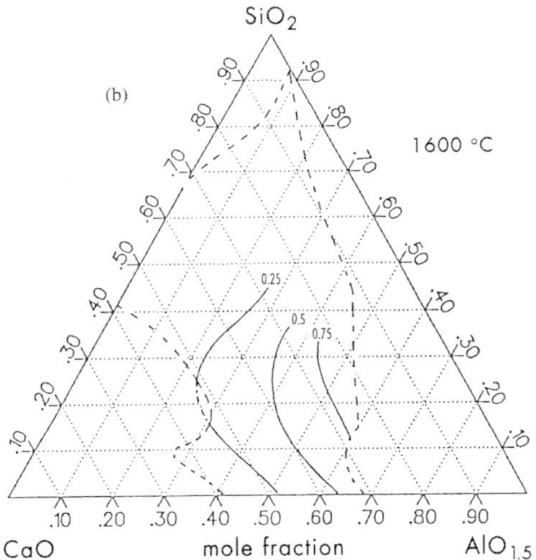

Figure 2.11: Activity of AlO$_{1.5}$ in CaO-SiO$_2$-AlO$_{1.5}$ slags at 1873K where standard state is pure solid AlO$_{1.5}$ with the relation 2 AlO$_{1.5}$ = Al$_2$O$_3$, and $(a_{AlO1.5})^2 = a_{Al2O3}$. [Ref. 7]

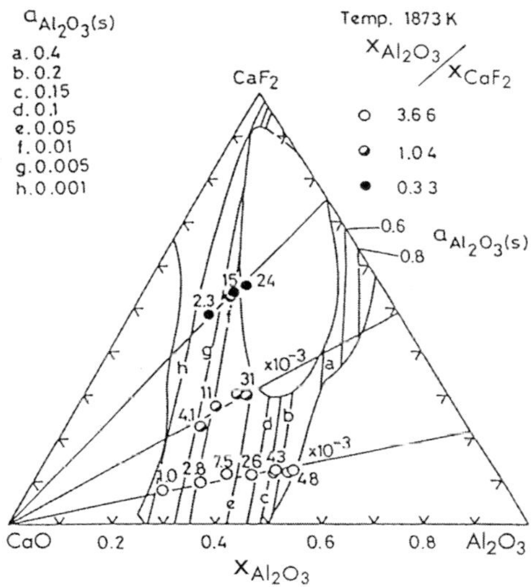

Figure 2.12: Activity of Al$_2$O$_3$ in CaO-CaF$_2$-Al$_2$O$_3$ slags at 1873K. Standard state is pure solid Al$_2$O$_3$. [Ref. 9]

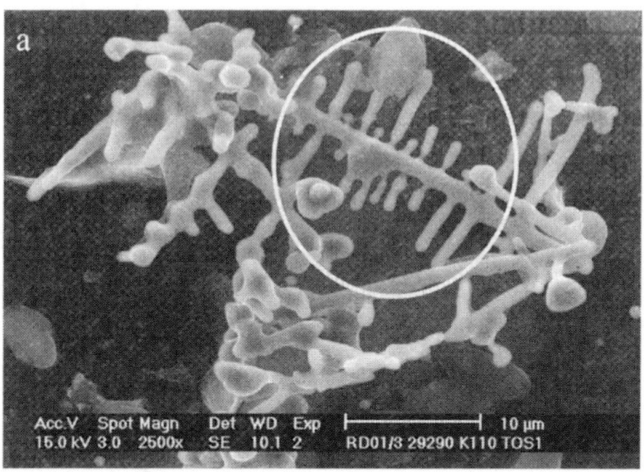

Figure 2.13: Coarsening of dendritic Al$_2$O$_3$ clusters formed while supersaturation of [Al] and [O] was high in LCAK steel (5 min after Al was added). [Ref. 10]

2.2.2.3 *Primary, secondary, and tertiary inclusions*

Indigenous inclusions precipitating when deoxidizing elements are added to the melt are called primary inclusions. They can be as large as exogenous inclusions, up to a few hundred micrometers. Indigenous inclusions subsequently precipitating due to the decrease in the solubility products while the melt is cooling are called secondary inclusions. They are usually smaller in size, from less than 1 μm up to roughly 20 μm, unless they precipitate on primary inclusions or agglomerate with each other. During the solidification, positive segregation of dissolved oxygen and deoxidizing elements occurs in the remaining melt when the solubility of the solutes in the solidified steel are lower than those in the melt, i.e., when the partition coefficient, k = C$_S$/C$_l$, of the solutes is less than unity. The positive segregation results in exceeding the solubility product, and finer inclusions precipitate in the remaining melt or grow on the suspended inclusions. During the solidification, the decreasing temperature of the solute enriched melt further decreases the solubility product, and additional finer inclusions can also precipitate or grow in the melt. To distinguish between the inclusions of the last two origins, researchers have referred to them as tertiary inclusions. For deoxidation with Al, the formation of tertiary inclusions is negligible since most of dissolved oxygen is used up by the time the mushy zone (the solid/melt coexisting zone between solid and melt) forms. For deoxidation with Mn and Si, however, both secondary and tertiary inclusion may form.

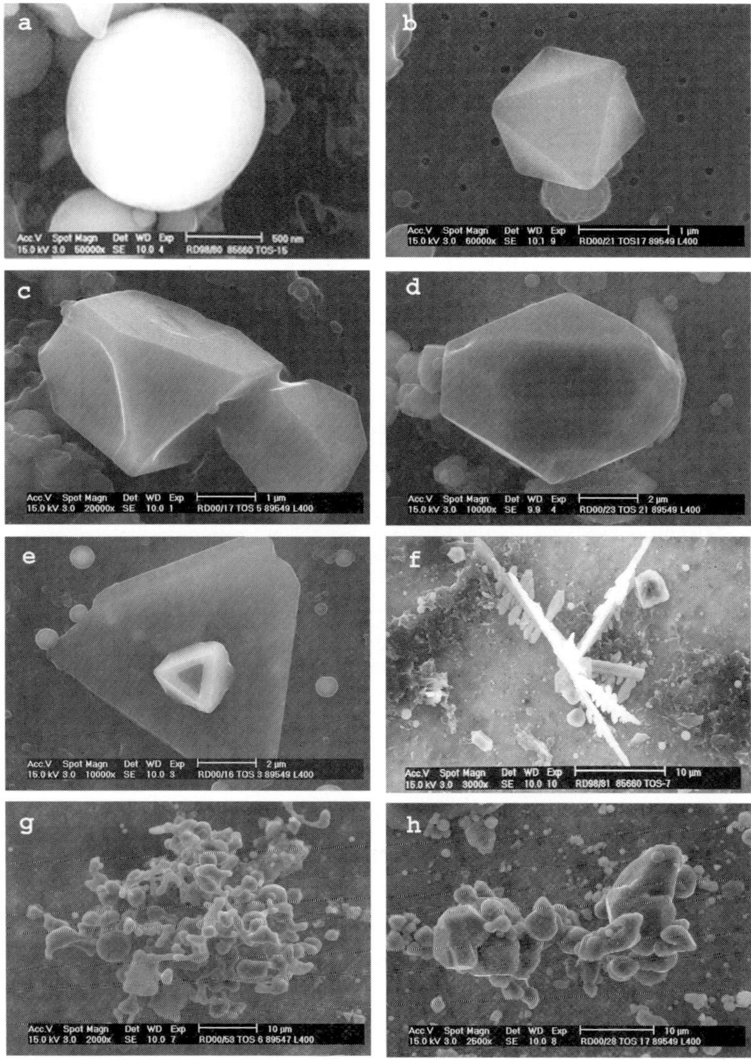

Figure 2.14: Typical Al$_2$O$_3$ inclusions in 300t LCAK steel after ladle refining with CaO-Al$_2$O$_3$ slag. (a) small spherical inclusions, (b) octahedral inclusions, (c) small polyhedral inclusions, (d) large polyhedral inclusions, (e) plate-like inclusions, (f) dendrite (g) cluster, and (h) aggregate. [Ref. 11]

Secondary inclusions can grow on the primary ones, and tertiary inclusions can grow on the secondary ones. Thus nucleation is not required for either of these inclusions to form. It is sometimes difficult to discriminate between the fine secondary inclusions and tertiary inclusions. Careful investigation of the morphology and chemistry can identify them in some cases. However, the secondary inclusions precipitated on the suspended small primary inclusions are often difficult to classify, and fine secondary inclusions held in the melt in the mushy zone are difficult to distinguish from the tertiary ones.

2.2.2.4 *Oxygen content of deoxidized steel*

Total oxygen content, $[O]_{total}$, in steel consists of oxygen dissolved in steel, $[O]_{dissolved}$, and oxygen contained in the primary, secondary, and tertiary oxide inclusions, $[O]_{inclusion}$, as given by

$$[O]_{total} = [O]_{dissolved} + [O]_{inclusion} = [O]_{dissolved} + [O]_{primary\ inclusion} + [O]_{sec\ ondary\ inclusion} + [O]_{tertiary\ inclusion}$$

$$(2.10)$$

2.2.2.5 *The deoxidation process*

The deoxidation process in the broad sense includes (1) decreasing dissolved oxygen in the melt with deoxidizing elements; (2) removing indigenous inclusions, mostly primary and some secondary ones, from the melt, while; (3) preventing the reoxidation of the melt and the pick up of exogenous inclusions in the melt. Strictly speaking, the deoxidation process should only include (1) and (2).

The grown/agglomerated inclusions float/rise to the surface and dissolve in the top slag, while some adhere to the refractory lining of the vessel. The balance, $[O]_{total}$ - $[O]_{dissolved}$, gives the amount of oxygen contained in the still suspended, fine-sized, alumina inclusions that are largely of indigenous origin, but some can be fine-sized exogenous inclusions. As bath stirring is continued and the inclusions are thoroughly removed, $[O]_{total}$ approaches to $[O]_{dissolved}$, provided that the reoxidation of liquid steel does not take place and supersaturation of

[O]$_{dissolved}$ is small enough to be disregarded. In turbulent steel melt that is stirred either by inert gas injection (Ar-Stirred Ladle, Ladle Furnace), by circulation with airlift pumping of Ar gas (RH process), or by electromagnetic means (ASEA-SKF), the supersaturation is usually considered small, which has been confirmed.

The values of [O]$_{total}$ and [O]$_{inclusion}$ depend very much on the practices of deoxidation and inclusion removal. The [O]$_{total}$ can be reduced by minimizing incoming exogenous inclusions into the deoxidized melt and by maximizing the indigenous inclusions being removed. Practical methods of achieving a low [O]$_{total}$ will be discussed later. In industrial practice, analysis of total oxygen content of Al-deoxidized 'clean steel' gives the best value of 3 ppm for high carbon steel melts, but may be as high as 10-20 ppm for medium and low carbon steel melts. The latter figures indicate that a large fraction of oxygen comes from still-suspended Al_2O_3 inclusions. For example, if all inclusion particles are assumed to be Al_2O_3 spheres of 1 µm radius, a melt of 1 cm^3 with 20 ppm total oxygen (4 ppm out of 20 ppm is dissolved oxygen for 0.04%Al) should include about 7 million Al_2O_3 particles. As long as the inclusions remain evenly distributed at such a small size, they do not impair most of the product properties in spite of rather high number density. The problem, however, is that they tend to collide, agglomerate, and form clusters that can become larger than the critical size, delivered to the mold, and finally get captured in the solidifying shell.

Although Al deoxidation is central to producing high quality steels for flat products, Si-Mn deoxidation is also important to produce some high quality steels for long products that do not permit the existence of sizable Al_2O_3 clusters since the clusters cause serious process problems. Steels for tire cord and valve springs fall in this category. For these steels, the minimization of inclusions and control of inclusions to keep them of a deformable composition (spessartite) during subsequent hot working are of vital importance. The minimization and the control have been done by Si-Mn-(Al) deoxidation under CaO-SiO_2-Al_2O_3 top slag with $CaO/SiO_2 \cong 0.8$ and lowest possible a_{SiO_2}. Technology for such inclusion control has been well developed as inclusion engineering.

2.3 Sizes and Shapes of Inclusions

2.3.1 *Size distribution of inclusions*

Sizes of inclusions retained in liquid steel after Al deoxidation and refining depend, as briefly discussed before, on the dissipation rate of stirring energy put into the melt, duration of the stirring, mode of stirring operation, etc. Under normal melt processing conditions where excessive entrainment is prevented, the maximum size of exogenous inclusions would be up to 100-200 μm. The size of indigenous inclusions can also be in the same range just after deoxidation, when the agglomeration and growth of primary inclusions are in progress. After the subsequent intense stirring operation during the refining process, however, the maximum size of exogenous and indigenous inclusions would decrease roughly to below 20 μm, as shown in Fig. 2.15 [12 and 38]. An exception is cluster-forming solid inclusions like Al_2O_3 particles. Elementary indigenous Al_2O_3 particles of smaller than 3 μm size form intermediate aggregates and loose clusters of small size. They collide and agglomerate to create large-sized loose clusters, which are subjected to densification by sintering, and get transformed to large dense clusters as shown by Yin *et al.* [13]. The majority of the large clusters are usually removed in the tundish. However, small-sized aggregates and clusters during the melt transfer from the tundish into the mold again form large clusters. This was actually observed by Kirihara *et al.* [14], when LCAK melt was refined by RH for more than 10 min and delivered to the tundish as clean melt without containing large clusters as shown in Fig. 2.16 (a). The melt, however, created large clusters well beyond 100 μm size in mold, but the large clusters were largely removed from the mold by flotation, and hence the resulting slab was as clean as the tundish melt as shown in Fig. 2.16 (b). Examples of the clusters/aggregates are seen in Fig. 2.14 (g), (h) and Fig. 2.17 after Dekkers [11]. The influence of the supersaturation on the shape of Al_2O_3 inclusions was extensively investigated by Steinmetz *et al.* [15].

The tendency of inclusion agglomeration is the main cause for nozzle clogging and defects occurring in Al-deoxidized steels. A simple way to prevent Al_2O_3 inclusions from forming a cluster is to decrease the number density of the elementary Al_2O_3 particles (and hence total [O] content) and to avoid turbulence of the melt during transfer to the mold, which would prevent their collision and agglomeration. In fact, cleaner Al-deoxidized steels containing less than 10 ppm $[O]_{total}$ decrease nozzle clogging and defects of alumina (Al_2O_3) origin to some extent, although it was not completely successful.

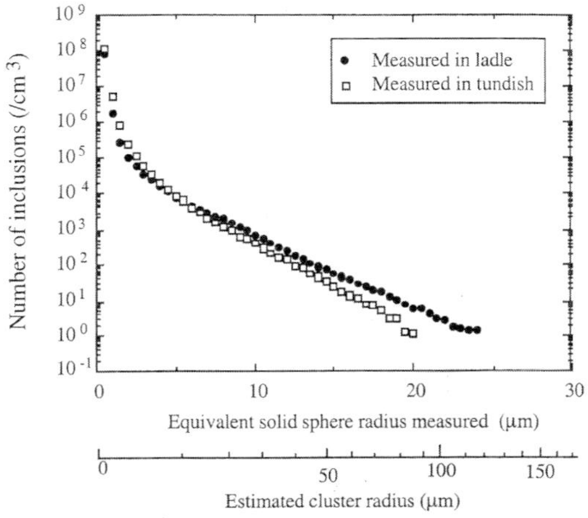

Figure 2.15: Measured Al_2O_3 inclusion size distribution in ladle and tundish in LCAK melt. [Refs. 12 and 38]

For a fully Al-deoxidized and reasonably refined steel melt with an average $[O]_{total}$ of about 20 ppm (including a few ppm of $[O]_{dissolved}$), the size distribution and number density of the sum of exogenous and indigenous inclusions are given roughly by

$$N(r) = N_0 \exp(-\varphi r) \tag{2.11}$$

where N_0 is the number density of the smallest discernible inclusions (~0.5 μm in diameter) in unit volume of the melt, r is the radius of inclusions, and φ is a constant. At $r = 0.5$ μm, N_0 is of the order of $10^6 \sim 10^7/cm^3$. This means that average inter-particle distance is of the order of 100 μm, which would allow inclusion particles to collide easily and agglomerate, and would yield larger inclusions under favorable circumstances. Unless they agglomerate into large clusters, the properties of steel are not appreciably degraded. The value of φ varies considerably for different operations, but N usually approaches near zero at 20 μm radius for a clean melt.

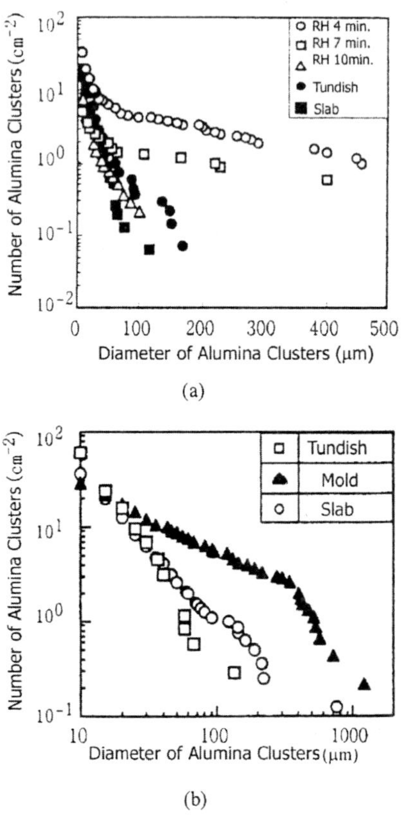

Figure 2.16: Change in size distribution of Al_2O_3 clusters in LCAK steel during RH processing and casting via tundish to slab (a), and formation of large agglomerates of Al_2O_3 clusters in CC mold (b). [Ref. 14]

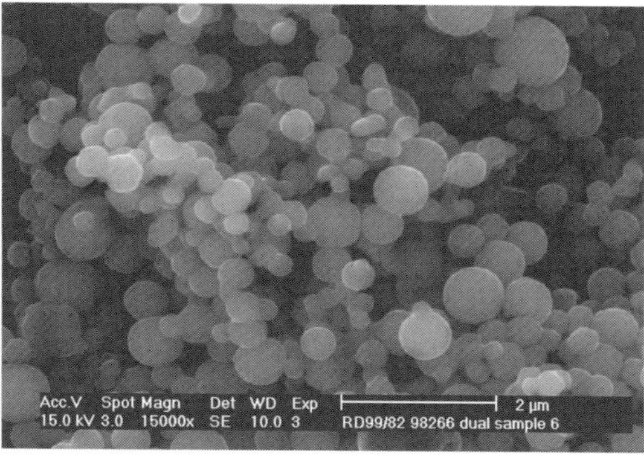

Figure 2.17: Al$_2$O$_3$ cluster consisting of smooth spherical constituent particles in LCAK steel melt. [Ref. 11]

2.3.2 *The shape of inclusions*

2.3.2.1 *Relative deformability of inclusions during hot rolling*

The shape of inclusions, both in molten and solid steel, depends very much on their melting (softening) temperatures, T_s. When the T_s is lower than the melt temperature, liquid inclusions deform into a globular shape as they are driven by the interfacial tension between the inclusions and steel melt. When solidified steel is hot rolled at temperatures near or above T_s, the inclusions elongate toward the rolling direction. An index of relative deformability, *R.D.*, of the inclusions in steel matrix is defined by Malkiewicz and Rudnik [16] as

$$R.D. = \omega \ln(\lambda) / \ln(h) \qquad (2.12)$$

where $\lambda = b/a$ is the ratio of length to thickness of an inclusion deformed by hot rolling the steel matrix, and $h = A_i/A_f$ is the ratio of cross sectional areas perpendicular to the rolling direction of the steel matrix before and after the hot rolling. Constant ω is 2/3 for rolling a square bloom or billet into a square product, i.e., equal transverse compressive strains normal to the rolling plane and in the rolling plane, whereas it is 1/2 for slab rolled to plate or strip product without width change, i.e., plain strain deformation. The value of R.D. is zero for non-deformable inclusions and unity for inclusions that elongate the same as a steel matrix.

2.3.2.2 *Classification of inclusions by shape after deformation of steel matrix*

Inclusions deform during hot and cold working of steel. A number of classifications of inclusions by shape after the deformation have been utilized. The following is often used as an example.

Inclusions classified as A-type are those that are largely homogeneous in chemistry, are glassy without precipitating a large volume fraction of hard crystalline phases before and during the hot working, and elongate during hot working. Inclusions with small hard crystals embedded in elongated glassy matrix are also classified as A-type. Typical examples are MnS and silicates with low melting temperatures.

Inclusions consisting of brittle and some plastic phases at hot rolling temperatures, are broken during hot rolling, leaving a discontinuous streak of broken fragments. Such inclusions are classified as B-type. The brittle phases in the inclusions can result from (1) phase separation during cooling of liquid inclusions, (2) composite structure formed as a consequence of the agglomeration of brittle- and plastic-particles, or (3) transformation of the surface of a brittle inclusion particle into a deformable crust. Aluminates and silicates with a softening temperature close to or higher than the hot rolling temperature belong to this class.

The third class of inclusions, which consists of agglomerated clusters of elementary solid particles in the steel melt, remains as clusters on solidification. Alumina clusters are typical examples of this class of inclusions. When hot rolled, the clusters disintegrate to form a stringer-like distribution of fragmented, smaller size agglomerates of the elementary particles in the rolling direction, if they do not deform at the hot rolling temperature. In an extreme case, the distance between the smaller agglomerates becomes so large that each agglomerate appears as if it is isolated under the limited view field of microscopic inspection. Such inclusions are classified as C-type inclusions. Solid non-deformable inclusions that form as isolated agglomerates from the beginning, such as irregular dense Al_2O_3 particles or globular SiO_2 particles, are also classified as C-type.

In steel melt, A-type inclusions and often many of the B-type inclusions exhibit a globular shape as mentioned before, whereas some of the B-type and most of C-type inclusions are in a massive irregular shape and cluster shape, respectively. Quite often, any combination of two or three out of the A-, B- and C-type inclusions may coexist as a composite.

The above classification is more or less arbitrary, since it is influenced to some extent by a combination of the reduction ratio and temperature of reduction of the steel matrix. In this regard, the effect of hot rolling temperature on the R.D. of manganese alumino-silicate inclusions in a 0.2%C steel, recorded by Maunder and Charles [17], is schematically shown in Fig. 2.18. Examples of the deformation of inclusions during hot rolling are schematically shown in Figs. 2.19 and 2.20 [17]. For more details of the relation between the deformability and microstructure, mineralogy, and chemistry of inclusions at various temperatures and reductions for hot rolling, readers are referred to a comprehensive work by Kiessling and Lange [18].

2.4 Influence of Inclusions on Steel Properties

Critical sizes and composition of inclusions often found in popular steel products are listed earlier in Table 2.1, where the LCAK cold rolled sheet undergoes a great reduction during hot- and cold-rolling of continuously cast slabs. Consequently, hard Al_2O_3 clusters smaller than 240 μm are broken into smaller fragments and do not cause any appreciable problem for conventional application of the sheet. However, a LCAK cold rolled thinner sheet for two piece DI cans undergoes, after blanking, more demanding drawing, ironing, flanging, and crimping. Cracks occur mostly on flanging and crimping of the cans. Occurrence of the defective cans exceeding 20~30 ppm (i.e. 20 to 30 cans out of 1 million cans) is an unacceptable limit by can producers. The limit may not be met unless the content and size of Al_2O_3 inclusions in the mother slabs are reduced to be less than about 30 ppm ($[O]_{total} \cong 15$ ppm) and 50 μm size.

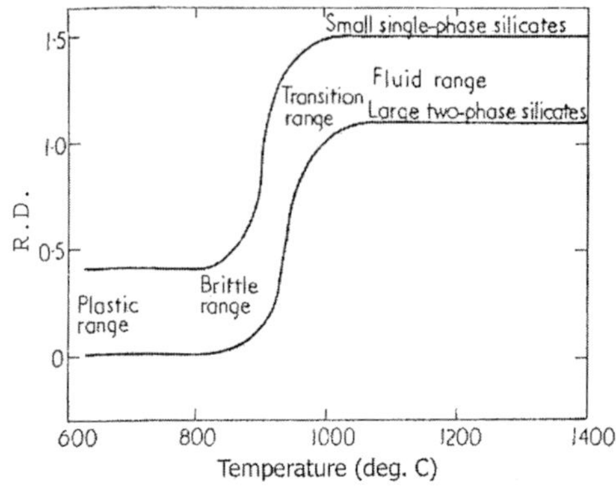

Figure 2.18: Schematic of relative deformability of silicates inclusions on hot rolling. [Ref. 17]

For UOE-pipe plate and ERW-pipe coil, inclusions larger than 200 μm and 150 μm, respectively, in the continuously cast slabs may cause defects in the final products that are detectable by non-destructive ultrasonic tests. These defects may become potential sites for fracture and weld defects. Inclusions larger than 100 μm in continuously cast blooms possibly result in cracks on cold forging of steel bars obtained by hot rolling of the blooms. Also, steel bars for tire cord would break during cold drawing if inclusions in mother cast bloom are bigger than 30 μm.

For more stringent applications, cleaner steels are required. Tolerable sizes of inclusions in more demanding steel products are listed earlier in Table 1.1 (Chapter 1). The table also lists the impurity concentration limits to avoid possible segregation of the impurities. The inclusion size limit for the plastic film laminated, cold rolled thin sheet for DI-cans is 5 μm, which is smaller than in the conventional DI-cans. In Table 1.1, sheet for the shadow mask in a cathode ray tube (CRT) and the lead frame for a large scale integrated circuit (LSI) may be judged defective if inclusions in the sheet are larger than 5 μm in size. These inclusions blur the edge of through holes on the shadow mask on etching, or cause cracks of the lead frame on press punching, resulting in a fuzzy image on the CRT or yield loss on punching.

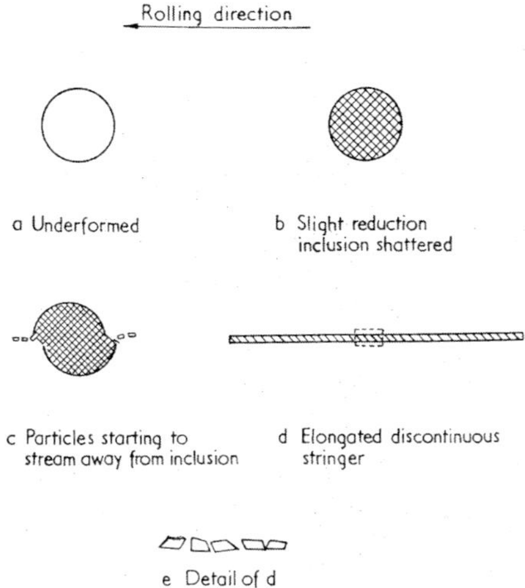

Figure 2.19: Stages of behavior of a glassy manganese alumino silicate inclusion in 0.2%C steel hot rolled in the very brittle temperature range. [Ref. 17]

In ball bearing steels, Ti is hazardous if it exceeds 20 ppm since the risk of the non-deformable TiO_2-TiN composite inclusions formation in the steel increases. The TiO_2-TiN inclusions may agglomerate to form large clusters (≥ 20 μm). Any other non-deformable solid inclusions larger than 15 μm would also appreciably decrease 10%-life to failure of the balls, rollers, and races of bearings since fatigue cracks form at such inclusions. To reduce the number density of these inclusions, $[O]_{inclusion}$ must be kept to less than 10 ppm, preferably 3 ppm, and [Ti] to less than 15~20 ppm. Non-deformable solid Al_2O_3 inclusions of 20 μm size in cast blooms cause rupture of high tensile strength wire and fiber for tire cord during cold drawing, and markedly reduce the wire productivity. To avoid these problems, Al_2O_3 inclusions should be kept to less than 20 μm size. Even more demanding criteria apply to valve spring wire.

It should be noted that the critical inclusion sizes depend on individual operational conditions and requirements for each product.

Accordingly, the critical sizes may not be taken as a universal standard but instead these are crude estimate for average operations.

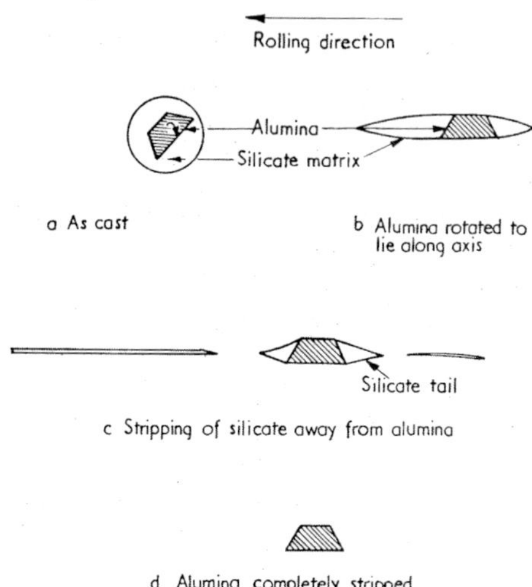

Figure 2.20: Stages of deformation in hot rolled 0.2%C steel of a liquid manganese alumino silicate inclusion containing one hard Al_2O_3 particle. [Ref. 17]

2.5 Measures to Reduce Inclusions

Inclusions of exogenous and indigenous origin in steel melt are removed from the melt by collision, agglomeration, and flotation. Large sized inclusions rise to the surface according to the Stokes' law, and are removed by dissolving in the top slag. However, these large inclusions which are capable to being floated by Stokes' rising velocity grow by Brownian, turbulent, and gradient collisions of much smaller-sized inclusions. Removal of the macro inclusions should be maximized and at the same time, pick-up of new macro inclusions should be minimized. The apparent rate, n, of inclusion removal per unit volume and per unit time is given by the difference between the rate of removal (n_r) and that of pick-up (n_p) as

$$n = n_r - n_p \tag{2.13}$$

Each term in Eq. (2.13) will be discussed in the following section.

2.5.1 *Maximizing inclusion removal in ladle refining*

The removal rate, n_r, of total oxygen content, $[O]_{total} = [O]_{dissolved} + [O]_{inclusion,}$ in the melt is superficially expressed as

$$n_r = K_o(C_t - C_\infty) \tag{2.14}$$

where K_o is the apparent mass transfer coefficient or the volumetric coefficient; C_t is the concentration of $[O]_{total}$ in the bulk melt at time, t; and C_∞ is terminal concentration of $[O]_{total}$ in the bulk, i.e., the sum of $[O]_{dissolved}$ in equilibrium with the deoxidizing element (C_e) and $[O]_{inclusion}$ in inclusions still suspended in the melt after sufficient period of refining.

From the thermodynamic point of view, $[O]_{dissolved}$ can be brought down to about 2.5 ppm at 1600°C, as discussed earlier, by deoxidizing a low carbon steel melt with $[Al] = 0.04$ mass% in a basic refractory lined vessel under a basic top slag for which $a_{Al_2O_3}$ is kept at about 0.25, and a_{SiO_2} and a_{FeO} are reduced to nearly zero. The concentration of inclusions suspended in the melt can be made sufficiently low, e.g. $[O]_{inclusion}$ of 8 ppm, by stirring the melt to enhance collision and agglomeration over time. If the value of C_∞ is reduced to a low of 10 ppm in the present case, and the rate of pick-up, n_p is also made insignificantly low, the removal rate, n_r is determined by the K_o term in Eq. (2.14). Under normal operation where the melts arc strongly stirred for refining, these particles collide and agglomerate by turbulent- and gradient-collision to form clusters. The clusters are subject to Stokes' flotation once they exceed a critical size, ascend to the melt surface, and get absorbed into the top slag or adhere to the melt/vessel refractory interface when they come into contact with the ladle or tundish wall.

Suspended inclusions, which are smaller than the critical size for Stokes' flotation, should grow by the collision and agglomeration. A function β_{ij}, which represents the collision frequency, is defined by

$$\beta_{ij} = N_{ij} / n_i n_j \tag{2.15}$$

where N_{ij} is the number of collisions per unit time and per unit volume of the melt between two spherical particles, i and j, with radius r_i and r_j, volume V_i and V_j, and number density n_i and n_j, respectively. Turbulent collision frequency is given by Saffman and Turner [19] with some modification by Higashitani *et al.* [20] as

$$\beta_{ij}^{turbulent} = 1.3\,\alpha\,R_c^3 \left(\frac{\varepsilon}{\nu}\right)^{\frac{1}{2}} \tag{2.16}$$

where $R_c = r_i + r_j$ is called the collision radius, ε is the dissipation of stirring energy per unit time and unit volume, and ν is the kinematic viscosity of the melt. Eq. (2.16) applies when R_c is much smaller than the Kolmogorov micro scale, $r_K = (\nu^3/\varepsilon)^{1/4}$, which is the minimum size eddy among the eddies with varying sizes in the turbulent flow. Eq. (2.16) indicates that turbulent collision frequency for the agglomeration of inclusions is proportional to the square root of the stirring energy dissipation rate, $\varepsilon^{1/2}$. In reality, stirring energy is applied to localized area of melt in a refining vessel. For simplicity, however, the stirring energy dissipation rate is often averaged over the total volume of the melt to estimate ε in Eq. (2.16).

For gradient collision with Stokes' flotation, Lindborg and Torssell [21] derived the following relation:

$$\beta_{ij}^{gradient} = \frac{2\pi\Delta\rho g}{9\eta}(r_i + r_j)^3 \, | \, r_i - r_j \, | \tag{2.17}$$

where $\Delta\rho$ is the difference in density between the particle (ρ_p) and the melt (ρ), and η is the viscosity of the melt. Brownian collision of the inclusions is known to be much less frequent than the turbulent and gradient one. Calculation by Taniguchi and Kikuchi [22] according to Eqs. (2.16) and (2.17) is given in Fig. 2.21. The figure shows that turbulent collision plays a major role in the agglomeration of small particles, whereas for larger particles gradient collisions are the dominant mechanism for agglomeration. The change in the number density of agglomerated particles consisting of a different number of elementary particles was also derived by the population balance equations. A similar approach was originally applied to steel deoxidation by Lindborg and Torssell [21], Linder [23], and Nakanishi and Szekely [24] to explore the relationship between the ε and the volumetric coefficient, K_o, for deoxidation.

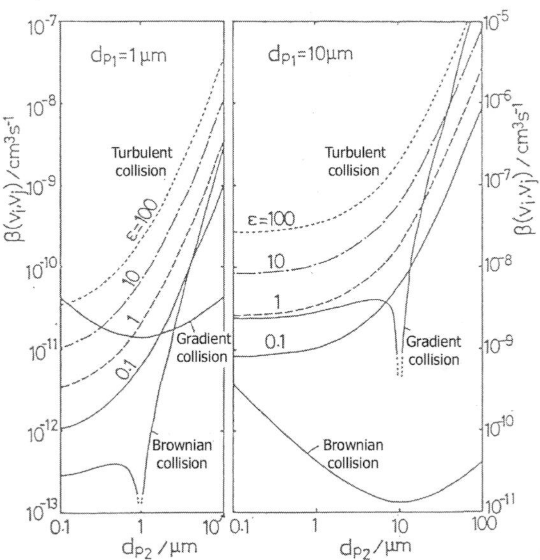

Figure 2.21: Frequency of collision between guest inclusion particles with various diameters (d_{p2}) and host inclusion particles with a small (d_{p1} = 1µm) or medium (d_{p1} = 10 µm) diameter. [Ref. 22]

In-plant observations inevitably include a negative contribution of the n_p term [entrainment term in Eq. (2.13) of top slag in the ladle] to the inclusion removal rate, and hence an accurate description of K_o is difficult to make at this moment. An empirical correlation indicated that K_o increased with ε to a different degree, depending on what type of metal bath stirring was used for ladle refining [Ref. 24]. Refining of steel melt is usually carried out in a ladle which is a cylindrical vessel. If a cylindrical metal bath of volume V is taken to be N perfectly mixed cells connected in series, then the amplitude, γ, of the decay curve of the concentration of solute, which is added pulse wise at any point in the bath, is given by Levenspiel [25] as

$$\gamma = 2\exp[-2\pi^2 \zeta^2 t / \tau] \qquad (2.18)$$

where τ is the time required to make the solute concentration homogenized within the total volume of the melt. The variance of impulse response in Eq. (2.18) is expressed by

$$\zeta^2 = \tau^2 / N \qquad (2.19)$$

The circulatory flow rate, Q, of the melt in the vessel is defined by

$$Q = V / \tau \qquad (2.20)$$

In the regime where either viscous forces, inertial forces or turbulent viscous forces dominate metal flow, and either convection or turbulent diffusion results in the dissipation of solute concentration, τ was related by Asai *et al.* [26] to the rate, ε, of stirring energy input per unit volume of the metal bath as

$$\tau \propto \varepsilon^{-n} L^{\delta} \qquad (2.21)$$

where L is the representative length of the vessel. Power factors $n = 1/2$ and $\delta = 0$ when viscous forces dominate melt flow and solute concentration dissipation is due to convection. They become, however, $n = 1/3$ and $\delta = 2/3$ or $n = 1/3$ and $\delta = 0$ when inertial forces or turbulent viscous forces dominate melt flow and the dissipation is by convection. When ε is relatively small, i.e., fluid flow is dominated by viscous force and the dissipation of the solute concentration is due to convection, τ is given by the following equation.

$$\tau \propto \varepsilon^{-1/2} \qquad (2.22)$$

The power factor in Eq. (2.22) is independent of the size of metal bath/vessel. It is to be noted that in a gas stirred vessel, ε (W/ton) is shown by Mori and Sano [27] to be proportional to gas injection rate, Q_g (Nm^3/min), blown from the bottom of the vessel into the steel bath:

$$\varepsilon = 6.18 Q_g \, T_1 \cdot \left\{ \ln\left(1 + \frac{h}{1.46 \times 10^{-5} P_2} \right) + \phi\left(1 - \frac{T_n}{T_1} \right) \right\} / M \qquad (2.23)$$

where T_1 (K) and h (m) are the temperature and height of the bath, P_2 (Pa) is the ambient pressure, ϕ is a coefficient, 0.06, T_n (K) is the temperature of the injected gas at the nozzle, and M (ton) is the weight of the bath. Originally, Sundberg [28] derived Eq. (2.23) with $\phi = 1$.

For more detailed discussion on the expression for ε, the reader is referred to a review by Mazumdar and Guthrie [29]. In normal metallurgical operations, the inertial/turbulent viscous force regime is utilized since it gives a much shorter τ, but sometimes operation in the viscous force regime is required, and hence Eqs. (2.21) to (2.23) become important.

Sakuraya *et al.* [30] summarized the empirical relationships for inclusion removal between K_o, Q_g, or ε, in industrial scale investigations in various refining vessels as

$$K_o \propto Q_g{}^n \propto \varepsilon^n \qquad (2.24)$$

As shown in Fig. 2.22 (see Fig. 8.8 in Chapter 8), the slope of the lines placed to cover the relationships gives n of about 0.4 on average.

The values of n observed by other researchers are in the range of 0.45-1.0 (Sandberg *et al.* [31], Fujii *et al.* [32]) under different operating conditions. In many cases, they fall within 0.40 to 0.59. This is close to the predicted value of 0.5. Values of K_o (min^{-1}) encompassed 0.04~0.12 for Ar bubbling in ladle at ε of 150~300 W/ton, 0.02~0.05 for RH at ε of 300~400 W/ton and 0.5 for upward stirring in ASEA-SKF at ε of about 120 W/ton. The values depended much on the mode of stirring, size of the melt, content of inclusions, degree of reoxidation by air, and top slag, glaze, and refractory of ladle. In Ar bubbling, K_o for the adhesion of Al_2O_3 on bubbles was found to increase with the injection rate of Ar, i.e., 0.21 with no gas injection, 0.51 at 5 Nl/min and 1.16 at 10 Nl/min at 1600°C (ex. ref. [30]).

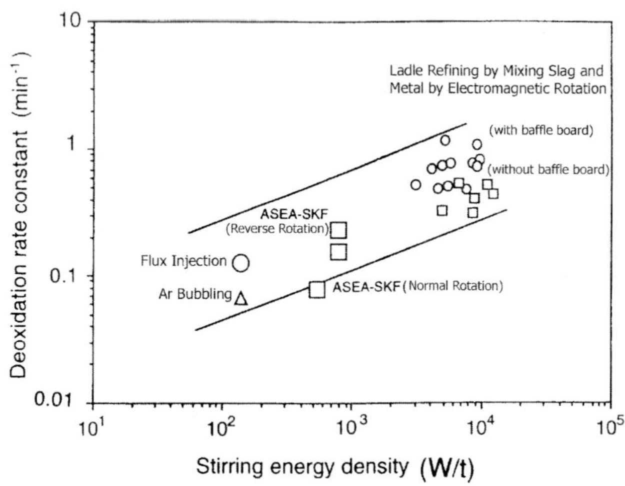

Figure 2.22: Apparent mass transfer coefficient in metallurgical vessels for oxygen /inclusion removal from deoxidized steel melt. [Ref. 30]

Although qualitative agreement between the theories and industrial observation is reasonable, the theories cannot be used quantitatively to design the refining vessels in view of significant variation in K_o at a given ε. The following factors are considered the reason for this variation.

First of all, input of inclusions into the melt during refining should be taken into account in industrial operation. It is a common understanding that the ratio of input to output of the inclusions in a refining vessel is related to the surface to volume ratio of the melt. The negative effect of stirring, caused by reoxidation by top slag and entrapment of top slag and vessel refractory, on the rate of inclusion removal, increases in smaller size melts. In contrast, positive effects on the removal rate by stirring are profound in larger size melts. Measures to remove inclusions by preventing the reoxidation and slag entrainment, and enhancing the stirring of the melt with LF, RH, and ASEA-SKF, are well developed and well documented in literature.

Second, the ratios of (1) agglomeration upon collision, (2) absorption into top slag, and (3) adhesion onto the refractory wall are either all taken to be unity (all successful), simply assumed to be in between 0 and 1 referring to industrial data, or taken to be fitting parameters. Further investigation on this aspect is needed to verify the theories.

Third, there is significant variation of the dissipation rate of stirring energy within the melt in a refining vessel. The variation exists, of course, between the processes and for each process. The theory applies to the situation where the stirring energy density is the same all over the melt. Accordingly, the application of the theory to design a refining process should be made carefully to implement the above factors in compliance with the characteristics of each process. Improvement on this issue was attempted by Miki *et al.* [33] for RH and Sheng *et al.* [34] for an Ar-stirred ladle.

Commercially popular equipments for removing inclusions during ladle refining (secondary refining) are depicted in Fig. 2.23. In all processes that incorporate gas stirring, increasing Q_g increases ε and K_o. On the other hand, excessive Q_g increases emulsification and entrainment of top slag and reoxidation of the melt splash or metal surface at the plume eye if injection is carried out in an oxidizing atmosphere. In such a case, careful optimization of Q_g is required. Also, in all processes, which employ top slag, the concentration and chemical potential of oxidizing components in the slag are to be minimized.

2.5.2 *Minimizing pickup of macro inclusions*

Macro inclusions occur mostly while the deoxidized and refined melt is transferred from the ladle to the tundish and from the tundish to the mold. They occur particularly during transient operation when exposure to air and the turbulence to entrain slag into the steel melt tend to happen. Measures to minimize the exposure and entrainment should be an integral part of a good tundish operation and these are discussed in Chapter 6.

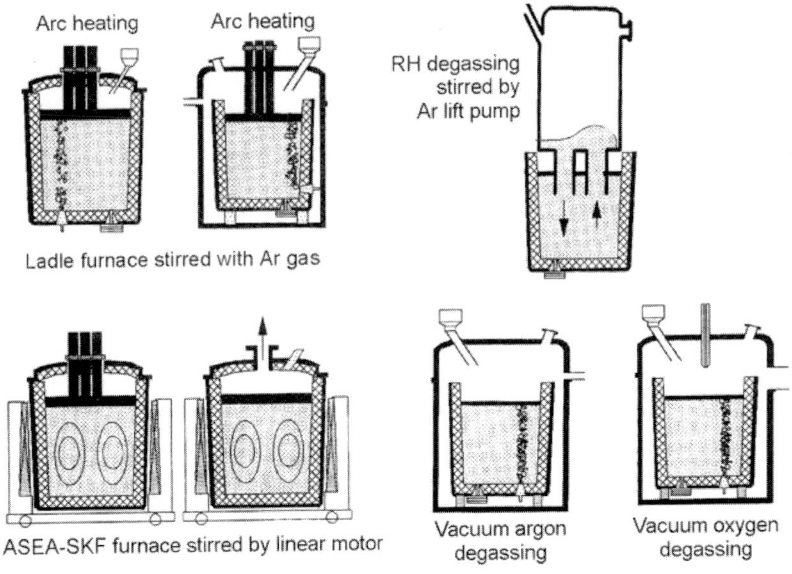

Figure 2.23: Typical equipment for secondary metallurgy to reduce impurity elements and inclusions in steel melt. [Ref. 1]

2.5.3 *Removal of macro inclusions during melt transport in tundish*

2.5.3.1 *The dominant mechanism of inclusion removal*

As stated earlier, the main objective of the tundish for inclusion removal is to let macro inclusions rise to the melt surface while the melt travels through the tundish into the mold. The dissipation rate of stirring energy density is low except for a small area where the ladle stream impinges into the melt bath in the tundish. The turbulent collision in the above area results in large spherical inclusions for liquid inclusions, large cluster inclusions for solid Al_2O_3, and spinel inclusions. In other areas in tundish, gradient collision prevails. As shown in Eq. (2.17), the gradient collision by Stokes' flotation is influenced by (1) specific density of inclusions that is low for liquid spherical silicate inclusions, high for solid Al_2O_3 inclusions, and higher for Al_2O_3 clusters that contain some fraction of steel melt; and by (2) a drag coefficient {see Eqs. (2.26) and (2.27)} that is small for the spherical shape inclusions, but appreciable for the irregular ones and the clusters. The upward component of the velocity of 3-dimensional melt flow in the tundish adds up to enhance the flotation velocity.

2.5.3.2 *Characteristics and modeling of the formation and removal of Al_2O_3 clusters*

Sawai *et al.* [35] developed a simulator for alumina inclusion removal in the tundish. A tundish was divided into finite space elements. Collision, agglomeration, and resulting change in size distribution of inclusions in each element were calculated according to Eqs. (2.15) ~ (2.17) and (2.34). Effects of the density and drag coefficient of the cluster morphology were not considered. The results were integrated with coupled diffusion transfer of the inclusions between the elements and removal of the inclusions by flotation to the melt surface. A parameter φ defined by $N_k = \varphi (N_i + N_j)$, $[0<\varphi<1]$ was introduced as a fitting parameter for successful agglomeration on collision between the inclusions. Another fitting parameter, ratio of inclusion entrainment, was also introduced that represented the fraction of inclusions that arrived at the surface but were entrained again into the melt without being removed by the surface slag. Calculated results showed that agglomeration by turbulent collision occurred only in the small area where the teeming stream of steel melt comes from the ladle into the tundish. The dominant mechanism of inclusion removal in the tundish was found to be gradient collision by Stokes' flotation. The parameter of successful collision of 0.1 (one out of 10 collisions forms agglomerate) and the ratio of entrainment of 0.7 (7 out of 10 inclusions return to the melt) made the best fit of the calculated results of inclusion removal, with the observation of samples taken from an industrial tundish. These values may be further confirmed by experimental validation in future.

Agglomerates of fine Al_2O_3 inclusion particles form coarse or dense dendrite-shaped or coral-shaped clusters with entrained steel melt in the clusters. Stokes' flotation velocity should therefore be modified for their apparent density, size, and drag coefficient. An interesting attempt was made by Tozawa *et al.* [36] to estimate them by fractal theory. Their modeling is presented here in some details. An example of Al_2O_3 cluster on a polished section of bomb specimen taken from the ladle melt immediately before teeming into the tundish is shown in Fig. 2.24. The alumina cluster diameter, D, in the melt and in solidified slab were plotted against the number, N, of constituent elementary Al_2O_3 particles of diameter, d, and is shown in Fig. 2.25. N is equal to $N_A^{3/2}$, where N_A is the number of the elementary particles (averaged 3 μm dia. each) observed on the polished surface. A solid line in the figure should give the relation between D and N instead of D and N_A, leading to

$$N = \chi D^{1.8} \tag{2.25}$$

where χ is equal to $d^{1.8}$ since D is equal to d when N is unity. This indicates that the fractal dimension of Al_2O_3 clusters is 1.8, which agrees well with the value of 1.78 ± 0.05 that was obtained by a 3-D computer simulation model of kinetic clustering for general clusters by Jullien *et al.* [37].

The drag coefficient, C_D, of model clusters of different degrees of agglomeration ($N_A \geq 10$, polyethylene elementary particles glued together) in water showed that

$$C_D = \frac{15}{Re} \; for \; Re \leq 3, \; or = \frac{8}{(Re)^{1/2}} \; for \; 3 < Re < 100 \tag{2.26}$$

where Re is the Reynolds number. The two C_D values are smaller than those for Stokes' law ($C_D = 24/Re$) and Allen's law ($C_D = 10/(Re)^{1/2}$) for spheres. Tozawa *et al.* reasoned that probably the circumambient part (the top part of the branches) of the clusters had fewer number density of the elementary particles and did not contribute to drag, and hence the effective D became smaller than the geometric D.

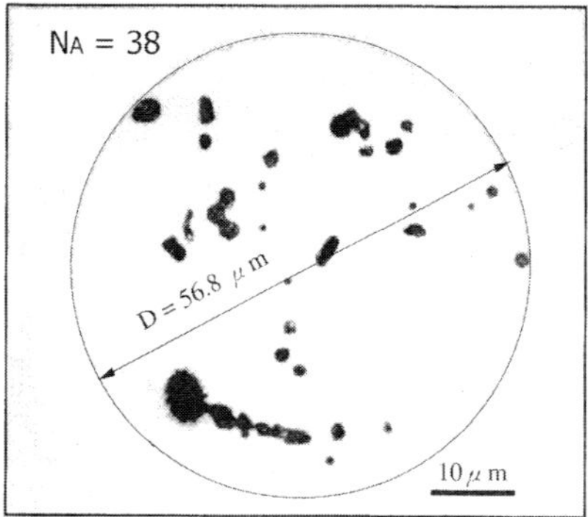

Figure 2.24: Alumina cluster revealed on polished section of a bomb sample taken from LCAK melt in ladle immediately before teeming into the tundish. [Ref. 36]

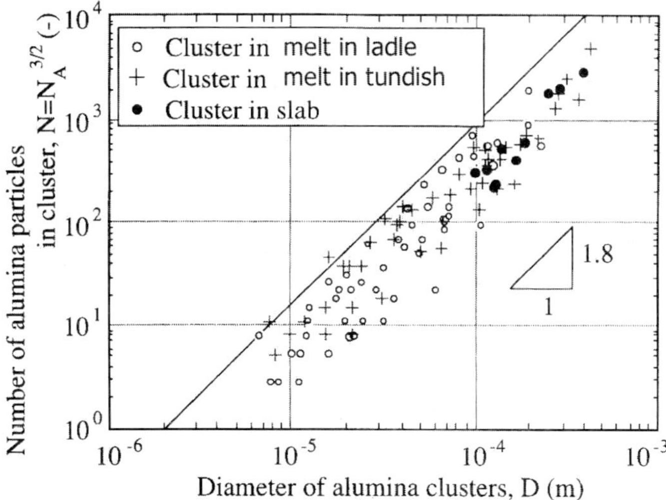

Figure 2.25: Distribution of number of constituent particles in alumina clusters observed in samples taken during teeming and casting of LCAK steel. [Ref. 36]

Terminal velocity, v, of the clusters is given by the balance of the buoyancy force and the drag force as

$$\frac{N\pi}{6}d^3(\rho - \rho_{Al_2O_3})g = C_D\left(\frac{\pi}{4}\right)D^2\left(\frac{1}{2}\rho v^2\right) \qquad (2.27)$$

One obtains values of v for clusters consisting of more than 10 elementary particles from Eqs. (2.26) and (2.27) as

$$v = \frac{8}{5}(D^{0.8}d^{1.2})(\rho - \rho_{Al_2O_3})\frac{g}{18\eta} \quad for \quad Re \le 3 \qquad (2.28)$$

or

$$v = (D^{0.2}d^{0.8})(\rho - \rho_{Al_2O_3})^{\frac{2}{3}}\left(\frac{g}{6}\right)^{\frac{2}{3}}(\rho\,\eta)^{-\frac{1}{3}} \quad for \quad 3 < Re < 100 \qquad (2.29)$$

where η is viscosity (Pa·s) of steel melt.

It was assumed that the space between the elementary alumina particles in the clusters is filled with steel melt of volume fraction f_{Fe}. Then, apparent density, ρ_c, of the clusters is given by

$$\rho_c = \rho\, f_{Fe} + \rho_{Al_2O_3}(1 - f_{Fe}) \qquad (2.30)$$

$$f_{Fe} = 1 - \left(\frac{d}{D}\right)^{3-D_f} \qquad (2.31)$$

The terminal velocity of cluster flotation was calculated from Eqs. (2-28)~(2.31) and compared with that of the spherical alumina according to Stokes' or Allen's law in Fig. 2.26. An Al_2O_3 cluster of 100 µm in diameter ascends at a speed of 1.2×10^{-3} m/s, which is only about 30% of the ascending velocity for a solid spherical Al_2O_3 inclusion of the same diameter. The ratio of terminal velocity gets smaller with the size of the cluster, indicating that large Al_2O_3 clusters are removed from the melt much slower than spherical ones of the same size.

For gradient collision, the collision frequency function becomes

$$\beta_{ij}^{gradient} = \pi(r_i + r_j)^2\,|v_i - v_j| \qquad (2.32)$$

In a tundish, Brownian collision frequency of fine inclusions is considered negligible for the formation of macro inclusions. The contribution to agglomeration comes mostly from gradient collision and some from turbulent collision as discussed before. The total collision frequency function is therefore given by

$$\beta_{ij} = \beta_{ij}^{gradient} + \beta_{ij}^{turbulent} \tag{2.33}$$

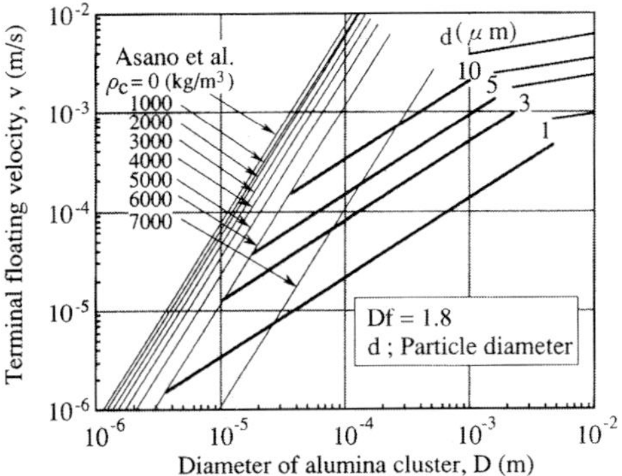

Figure 2.26: Terminal flotation velocity of alumina clusters according to Stokes' law (thin lines, Asano *et al.*) and present calculation by Eq. (2.26) (thick lines). [Ref. 36]

When a cluster consisting of *i* elementary particles (clustering degree *i*, abbreviated as cluster i hereafter) and a cluster j collide to form a cluster *k* (= *i* + *j*), the net rate of the formation of cluster k is given by use of population balance equation derived by Smoluchouski as

$$\frac{dn_k}{dt} = \left(\frac{1}{2}\right) \sum_{i+j=k} N_{ij} - \sum_{i=1}^{\infty} N_{ik} = \left(\frac{1}{2}\right) \sum_{i+j=k} \beta_{ij} n_i n_j - n_k \sum_{i=1}^{\infty} \beta_{ik} n_i \tag{2.34}$$

where N_{ij} is the number of collisions between clusters *i* and *j* in unit time, and n_i is the number concentration of cluster *i*.

The concentration of cluster *i* (C_i in mass %) in the melt was given approximately by

$$C_i = 100\, i \left(\frac{4\pi}{3} r^3 \right) \left(\frac{\rho_{Al_2O_3}}{\rho_{Fe}} \right) n_i \qquad (2.35)$$

with the density, ρ in kg/m^3, of subscript materials. Eqs. (2.33) and (2.34) gave the rate of evolution, $f(C_k)$, of cluster k as

$$f(C_k) = \left(\frac{3\rho \cdot 10^{-4}}{4\pi\rho_{Al_2O_3}} \right) \left(\frac{k}{r^3} \right) \left\{ \frac{1}{2} \sum_{i+j=k} \beta_{ij} \left(\frac{C_i}{i} \right) \left(\frac{C_j}{j} \right) - \left(\frac{C_k}{k} \right) \sum_{i=1}^{\infty} \beta_{ik} \left(\frac{C_i}{i} \right) \right\} \qquad (2.36)$$

In a tundish, Stokes' flotation is influenced by 3-D flow velocity of steel melt. Governing equations of the flow include (1) the equation of continuity, (2) the equation of momentum balance, and (3) the equation of enthalpy balance. These equations together with (1) the k-ε model for turbulent flow, (2) the iteration of cluster concentration and terminal velocity of flotation for each degree of agglomeration, and (3) an approximation that effective Schmidt number and effective Prandtl number for the melt are both unity gave, after some manipulation, the concentration equation for cluster i as

$$\frac{\partial(\rho C_i)}{\partial t} + \frac{\partial(\rho u C_i)}{\partial X_j} = \frac{\partial}{\partial X_j} \left(\rho D_{eff} \frac{\partial C_i}{\partial X_j} \right) + \rho f(C_i), i = 1 \sim |\, i_{max}\,| \qquad (2.37)$$

with

$$\rho k_{eff} = \rho D_{eff} = \eta_{eff} \qquad (2.38)$$

Here, $u = u\,(u_1,\, u_2,\, u_3 + v_i)$ is the flow velocity of a cluster i, for which terminal velocity of flotation is v_i (m/s), $\eta_{eff} = \eta + \eta_t$ is effective viscosity of the melt (Pa·s), and η_t is turbulent viscosity.

The second term on the right hand side of Eq. (2.37) was given by Eq. (2.36) as

$$f(C_i) = \left(\frac{3\rho \cdot 10^{-4}}{4\pi\rho_{Al_2O_3}} \right) \left(\frac{i}{r^3} \right) \left\{ \frac{1}{2} \sum_{j+k=i} \beta_{jk} \left(\frac{C_j}{j} \right) \left(\frac{C_k}{k} \right) - \left(\frac{C_i}{i} \right) \sum_{j=1}^{i\,max} \beta_{ji} \left(\frac{C_j}{j} \right) \right\} \qquad (2.39)$$

for $i = 1 \sim i_{max} - 1$, and

$$f(C_i) = \left(\frac{3\rho_{Fe}}{4\pi\rho_{Al_2O_3}}\right)\left(\frac{i}{r^3}\right)\left\{\frac{1}{2}\sum_{j+k \geq i\,\max} \beta_{jk}\left(\frac{C_j}{j}\right)\left(\frac{C_k}{k}\right)\right\} \qquad (2.40)$$

for $i = i_{max}$.

One half of a 70-ton tundish for a two-strand slab caster (boat type, length 3.85 m, width 1.79 m max, depth 1.0 m max.) was divided into 26 x 46 x 20 grid cells for computational mesh. Melt flow velocity was calculated for a throughput rate of 4.1 ton/min. In the calculation, observed heat fluxes of 5.24 and 11.27 kcal/m^2.s were used for the tundish wall and free surface. The free surface was taken to be a frictionless impervious boundary. Two dimensional velocity profiles on the center plane of the tundish are shown in Fig. 2.27.

The distribution of clusters with i in the range of $1 \sim 29$ was calculated for ULCAK melt containing initial $[O]_{total}$ of 50 ppm. The size and number density of the clusters were represented by Eq. (2.11). The clusters were assumed not to adhere to the wall. Also, it was taken as a fitting parameter that 80% of the cluster flux ($C_i\, v_i$) that reached the free surface was to be removed from the melt, while the balance was entrained back into the melt. The size distribution of clusters was separately observed on bomb samples taken in the bath at 400 mm depth near the inlet point (long nozzle) and at 560 mm depth near the outlet point (tundish nozzle) of the tundish. The calculated results are in reasonable agreement with the observed results, as shown in Fig. 2.28. Clusters i in the range of $i = 5\sim90$ decreased substantially from the inlet to outlet during the melt flow in the tundish. For clusters i greater than 90, such a trend is not clear since number density of those large clusters is too small to tell the difference. The anomalous jump of the calculated curve at the inlet for $i \geq 29$ was considered to have resulted from the fact that the collision and agglomeration of clusters i of 29 and clusters i greater than 29 were ignored in the calculation.

2.5.3.3 *Influence of flow and temperature of melt on the removal of alumina clusters*

Melt flow in the tundish should have strong influence on the removal of macro inclusions that are subject to gradient collision and agglomeration followed by Stokes' flotation. A critical issue is to produce a sufficient upward velocity component over a large fraction of space in the tundish. For this objective, the tundish is often divided into two compartments, an inlet- and outlet-compartment, by installing a baffle board, weir and dam combination, or similar combinations. Details of the design and function of the furniture will be discussed in Chapter 6. The flow is also influenced by non-isotropic distribution of the melt temperature in the tundish, i.e., steel melt flows from the hot inlet compartment into the cooler outlet compartment, while the melt in contact with the free surface and wall refractory is being cooled. The influence of such temperature gradient on the flow will be discussed in detail in Chapter 5. Convective flow (thermal buoyancy) caused by the non-isothermal situation causes additional modification of the flow partly controlled by the furniture. The convection can change the 3-dimensional characteristics of the flow significantly.

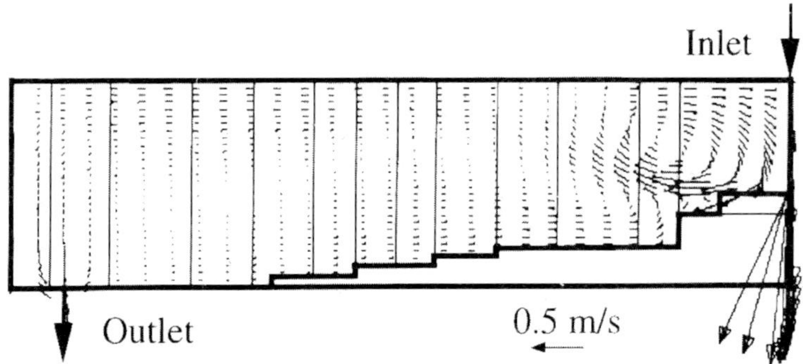

Figure 2.27: Calculated velocity profiles at longitudinal center plane of 70-ton tundish when steel melt is fed at 4.1t/min. [Ref. 36]

Another interesting 3-D model simulation was developed in this regard by Miki *et al.* [38], for melt flow and temperature evolution coupled with inclusion removal in a tundish in both steady and transient conditions. For turbulent melt flow, a k-ε model was employed. Trajectories of inclusions were calculated through the flow field. Evolution of inclusion size distribution was obtained for the average flow and turbulence conditions similar to those described in the preceding section. The time evolution of a general inclusion concentration field was calculated with a scalar diffusion model. The rest of the chapter follows their modeling.

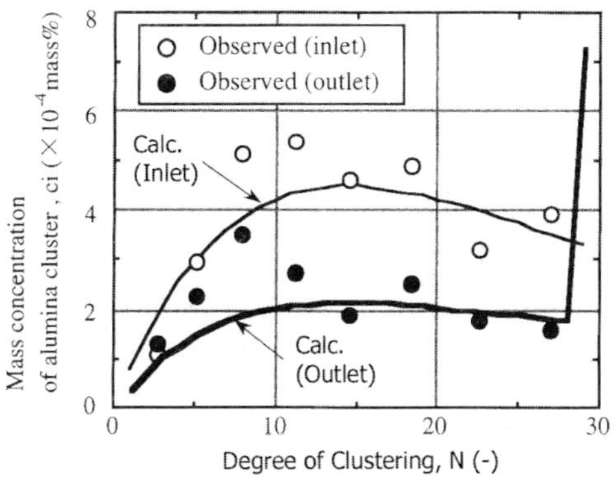

Figure 2.28: Comparison of calculated and observed size distribution of alumina clusters in LCAK steel melt in a 70-ton tundish. [Ref. 36]

For melt flow and heat transfer in the 3-D domain of tundish, the 3-D continuum transport model solved the mass conservation equation and three momentum balance equations for the velocities, u_i, and pressure, P.

These equations were solved for a 30 ton tundish (about 4m length x 1m width) into which steel melt of 1853 K was poured to a depth of 1.1m at a fixed rate of 1.7 ton/min. Here, the inlet velocity was 0.43 m/s, turbulent kinetic energy was 0.003 m^2/s^2, turbulence dissipation rate was 0.0003 m^2/s^3, and the residence time was 1060 s.

The steady distribution of melt flow and temperature at the center plane of the tundish was calculated with the coupled flow-temperature model. The through hole at the bottom of the dam was 0.3 m square. Heat fluxes from the top surface, bottom surface, and side walls were taken from Chakraborty and Sahai [39] to be 15, 1.4 and 3.2-3.8 kW/m^2, respectively.

For an isothermal case, the melt was found to form a short circuit flow along the bottom of the tundish. In a non-isothermal case, however, melt flow from the inlet compartment is, as shown in Fig. 2.29, directed upward by buoyancy. The flow travels along the surface, is deflected downward by the wall end, and partly escapes from the tundish. However, the rest is delivered back along the bottom, where it collides with the flow from the through hole and circulates to the surface again.

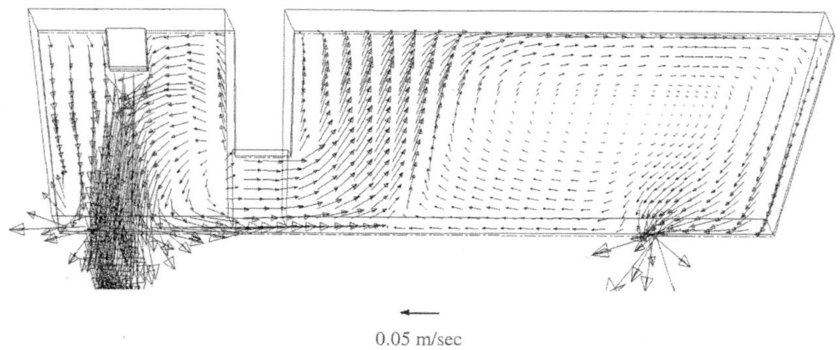

Figure 2.29: Steady flow distribution of steel melt at the longitudinal center plane calculated with a coupled flow-temperature model. [Ref. 38]

The inclusion trajectory was calculated using a Lagrangian particle tracking method, which solved a transport equation for each inclusion as it traveled through a numerically calculated steady state melt flow field. The mean local inclusion velocity components, $\overline{u_{ci}}$, were obtained from the balance of drag and buoyancy forces as

$$\frac{\partial}{\partial t}\overline{u_{ci}} = \frac{18\,\eta_0\, C_D\, \text{Re}}{24\,\rho_c\, D_c^2}\left(u_i - \overline{u_{ci}}\right) + \left(\frac{\rho_c - \rho}{\rho_c}\right)g_i \qquad (2.41)$$

A discrete random walk model was applied during most inclusion trajectory calculations to simulate the chaotic effect of the turbulent eddies on the inclusion paths. A random velocity vector, u_i', was added to the calculated time averaged vector, $\overline{u_{ci}}$, to obtain the inclusion velocity, $u_{ci,}$ at each time-step as it traveled through the melt. Each random component of the inclusion velocity was proportional to the local turbulent kinetic energy level, K, according to

$$u'_{ci} = \zeta_i \left(\overline{u_i'^2} \right)^{\frac{1}{2}} = \zeta_I \left(\frac{2K}{3} \right)^{\frac{1}{2}} \qquad (2.42)$$

where ζ_i is a random number distributed between -1 and 1 that changes at each time step. Inclusions were injected computationally at different locations distributed homogenously at the inlet. Each trajectory was calculated through the steady state flow field until the inclusion either was trapped on the top surface or exited the tundish outlet. Inclusions touching a side wall were assumed to be reflected back to melt.

In calculating the trajectory, the circumambient diameter of alumina clusters was taken to be a fitting parameter that made the calculated results agree with the observations. The circumambient diameter fell between D_c defined by fractal theory and equivalent solid alumina sphere with the same mass as the cluster. An example of the calculated trajectory of a 50 μm cluster in non-isothermal melt flow (Fig. 2.29) is given in Fig. 2.30. It shows that the complex recirculation in the inlet compartment and rising melt flow in the outlet compartment help inclusions to be removed to the top surface.

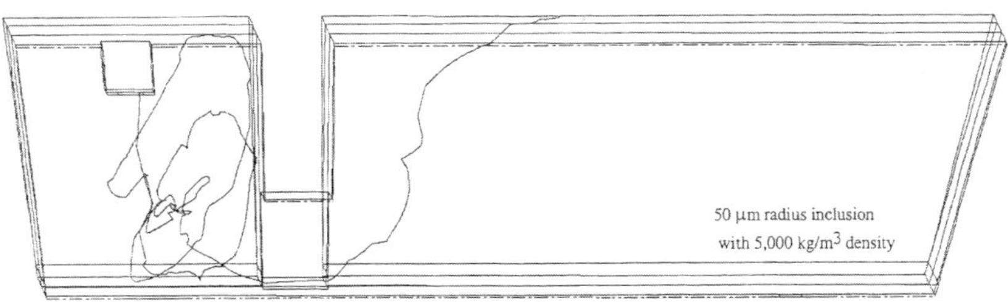

Figure 2.30: Typical inclusion trajectory calculated using random walk model in steady non-isothermal flow field of steel melt. [Ref. 38]

For each cluster size, trajectories of 200 clusters were calculated, and the ratio of inclusion removed was plotted in Fig. 2.31. The removal ratio increased with the cluster size. The ratio was greater for a non-isothermal case, where buoyancy further helped flotation of the smaller clusters. The removal ratio of clusters was similar between the inlet and outlet compartment, but was somewhat greater in the inlet compartment for large clusters, as shown in Fig. 2.32.

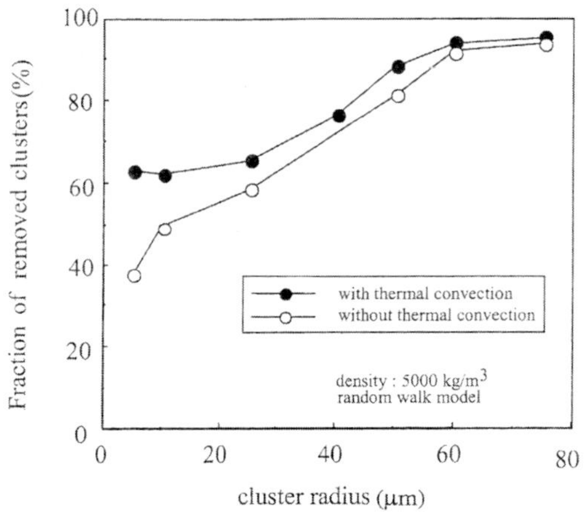

Figure 2.31: Calculated fraction of clusters removed in tundish for isothermal and non-isothermal models. [Ref. 38]

Inclusion removal was also evaluated by combining collision and agglomeration in turbulent eddies and during Stokes' flotation with the 3-dimensional melt flow. The rate of inclusion removal to the free surface (S) and the rate of the formation of fine inclusions by reoxidation (G) were included as

$$\frac{d}{dt}f(r_k) = \frac{1}{2}\sum_{i=1}^{i=k-1} f(r_i)f(r_{k-i})\beta_{i,k-i} - \sum_{i=1}^{i\max} f(r_i)f(r_k)\beta_{i,k} - S + G \quad (2.43)$$

under the condition $r^3_{k-i} = r^3_k - r^3_i$. Here, G was assumed to be 2.5 x 10^9 of 1 μm dia. inclusions/m³/s. β was the same as given in Eqs. (2.15)-(2.17), and Eq. (2.43) compares with Eqs. (2.33) and (2.34). Based on the 3-D melt flow model, a mean turbulent energy dissipation rate of 0.0004 m²/s³ was assumed.

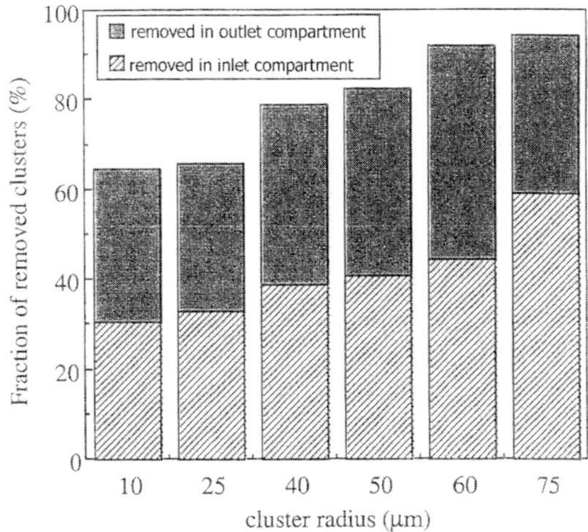

Figure 2.32: Calculated fraction of the inclusions removed in the inlet and outlet compartments of a 30-ton tundish. [Ref. 38]

The calculated results of the inclusion removal with a calibrated lumped model that included flotation, collision, and reoxidation are shown in Fig. 2.33 together with the result of the random walk inclusion trajectory model (flotation only, constant cluster size with assumed density of 5000 kg/m^3). Measured results of inclusion removal on samples taken over the outlet port at 500 mm depth from the free surface in the 30-ton tundish are also included in Fig. 2.33. The calibrated lumped model gave better agreement with the measurements than the random walk inclusion trajectory model in spite of many assumptions involved. These models were also applied to evaluate the inclusion removal during the transient period of tundish operation with some success.

2.5.4 *Tundish design and inclusion removal*

2.5.4.1 *Design criteria for inclusion removal*

The tundish is located just above the CC mold. Any disturbance that causes reoxidation and slag/flux entrainment results in the formation of macro inclusions. These could then be carried over into the mold and may result as casting defects. In addition, the space available between the ladle and tundish is usually limited. Accordingly, a tundish is not equipped with refining and alloying facilities except for an addition of micro alloying elements or calcium to reduce nozzle clogging, if at all necessary. The goals of a tundish are to minimize heat loss, deliver the melt evenly into molds, minimize the formation of macro inclusions, and maximize their removal.

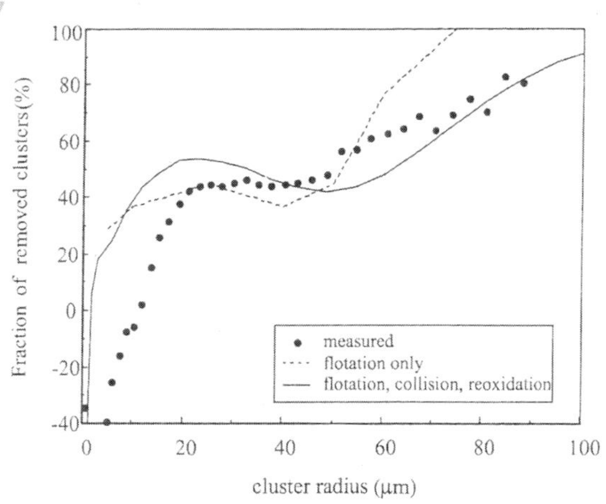

Figure 2.33: Comparison of observed and calculated fractions of inclusions removed in a 30-ton tundish. [Ref. 38]

The cost issue is another important factor to consider. As mentioned later in Chapter 8, the hot cycle of the tundish is practiced to reduce the tundish refractory consumption drastically. The complicated structure of a tundish with furniture (flow control devices) is not suitable for a hot cycle tundish. Also, the tundish needs to be exchanged on the fly to keep up with the casting of many heats in sequence. Durability of the refractory is mandatory and a very important factor. The tundish shell needs to be sturdy, which should not allow any deformation. Even a small deformation may cause off-centering of the SEN in the mold, and may result in asymmetric melt flow in the mold that adversely affects inclusion removal and enhances mold flux entrainment.

Some of the above mentioned issues contradict each other in designing of a tundish. It is, therefore, necessary to make a good compromise among them to reduce macro inclusions. The following factors need particular attention in making a tundish effective for reducing macro inclusions:

#1. Steady state period of casting requires:

(1)   A reasonably shallow and wide section in the tundish to promote Stokes' flotation;

(2)   Buoyant melt flow from inlet toward the outlet section to assist the flotation;

(3)   Large volume fraction of buoyant flow in the first half of the tundish length;

(4)   Melt flow with less turbulence along the tundish flux/melt interface;

(5)   No short circuiting of melt flow from the inlet section to the exit (tundish nozzle);

(6)   Minimal dead volume for the melt in the tundish;

(7)   Sufficient residence time for the melt to promote flotation;

(8)   An argon gas shrouding pipe or long nozzle for ladle melt discharge into the tundish;

(9)   Thermal insulation and protection against reoxidation by argon gas injection with lid or a tundish flux cover;

#2. Non-steady state period of casting requires:

(10)   Prevention of slag carry over by vortexing and draining from the ladle to the tundish;

(11)   The above slag carry over issue also applies to tundish to mold transfer. This is a constraint in making a tundish excessively shallow;

(12)   Suppression of turbulence caused by impinging melt stream to the tundish at the ladle opening;

(13)   Sustaining an inert atmosphere at ladle opening and ladle change;

(14)   Active compensation for temperature drop.

Implementation of these factors depends much on local circumstances. Each casting plant has developed their optimized preventive operational schemes. Some typical examples that have proven effective are mentioned to some detail in Chapter 6. Among them, one of the most important factors of tundish design for inclusion removal has been the dimension and profile of the tundish.

2.5.4.2 *Effect of dimensions and profile of the tundish on inclusion removal*

Development of modern slab casters for mass production of quality steel in an integrated steel plant has been toward high throughput rate, high speed sequence casting for productivity, and cost considerations. With such a slab caster, it would become possible to make an ideal combination of 1 large BOF - 1 LF/RH - 1 two-strand slab caster - 1 hot strip mill, that can be operated at a minimum investment and low running cost.

Under the circumstances where steel melt is well refined and transferred into a tundish, current tundish technology (mentioned in Chapters 6 and 8) is capable of delivering the melt into the mold without contamination by macro inclusions so far as steady state casting is concerned. What matters is the contamination by macro inclusions that inevitably occurs during non-steady state period of casting when opening and emptying of ladles have to be done. One source of such contamination is the size of the tundish. If the tundish is smaller relative to a high throughput rate, it will suffer from quick decrease of bath depth, which may cause vortexing at tundish exits during the ladle change period. The decrease in melt depth causes detrimental carry over of tundish slag into the mold as macro inclusions. Another source is that during the non-steady state operation, the Ar-shield of the melt supply from ladle is often broken, allowing splash or plume of the melt being reoxidized. Also, the floating ladle slag and tundish slag could be pushed into the melt. Measures to minimize such turbulence exist, but have shown limited success. A small tundish does not offer enough residence time for the removal of the inclusions once they are introduced. Thus, tundish design should emphasize the negation of the bad influences of the above factors.

To counteract these problems, the volume and depth of melt in a tundish were increased with caster productivity as summarized by Marique [40] in Figs. 2.34 and 2.35. For 10 years after this study, even larger tundishes in excess of 70 tons were rather commonly used for highly productive casters. An example of the decrease of macro inclusion at the ladle change when tundish capacity was increased from 65 tons to 85 tons is briefly shown in Fig. 2.36 [41]. The increase made it possible to raise the retained melt volume at ladle change from 35 tons to 55 tons. Also, the melt residence time and bath depth during steady state casting increased from about 6 min to 8 min and 1200 mm to 1400 mm, respectively. Significant decrease of macro inclusions at the non-steady state is obvious for the large tundish.

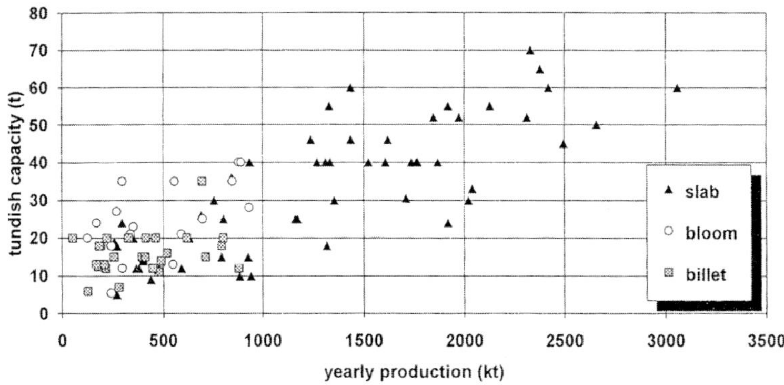

Figure 2.34: Tundish capacity increased with yearly production of European casters. [Ref. 40]

The effect of the ratio of width to depth of the tundish also needs some discussion. For Stokes' flotation, a shallow profile would be better.

In a model computation by Nakaoka *et al.* [42] shown in Fig. 2.37 (a), a deep tundish with a width to depth ratio (W/D) of 0.44 exhibited short circuit flow along the side walls to the tundish outlet. These results agreed well with a 1/3 scale water model observations. The short circuit flow increased the fraction of macro inclusions carried from the tundish into the mold, whereas it was much less in a square section tundish (W/D = 1) of the same volume; see Fig. 2.37(b). Such dependence on W/D was, however, marked for a small volume tundish only. The dependence was much less for a large volume tundish as shown by water model experiments in Fig. 2.38 in the range of W/D 1.0-2.0 (compare curves for 0.12 m^3 and 0.41 m^3).

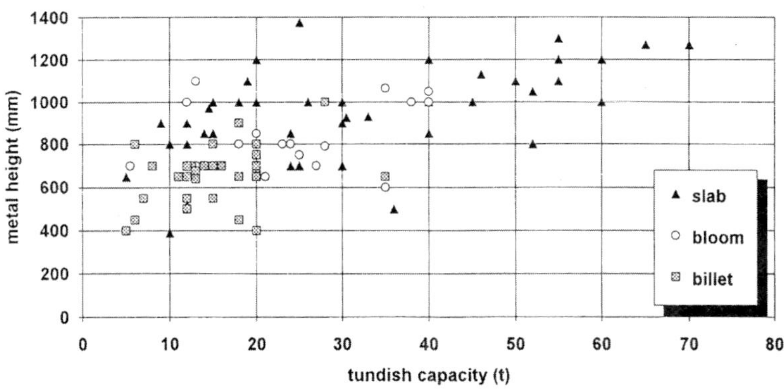

Figure 2.35: Tundish bath depth increased with tundish capacity of European casters. [Ref. 40]

The model computation and water model experiment mentioned above did not include the influence of the temperature gradient. The water model experiment did not use a full scale model. Thus, the results would be subject to some modification by a more precise computation and water model experiments. However, the trend that the large tundish is insensitive to its W/D ratio for the removal of macro inclusions, has been confirmed in industrial operations, validating the move toward the enlargement of the tundish with sufficient bath depth to avoid vortexing. In fact, increasing the W/D ratio is not always beneficial for a large tundish, i.e. heat loss increases more, shell structure of tundish becomes susceptible to deformation, and the shallow melt depth is risky for tundish slag entrainment as mentioned earlier. A minor gain in the decreased outflow rate of macro inclusions does not rationalize designing a tundish with a large W/D ratio at the risk of the above disadvantages. In industrial tundish design, therefore, a compromise has been made among these factors, the end result being a depth of around 1.1-1.4 m, W/D ≤ 1, and a melt weight of 50-80 ton for a highly productive slab caster.

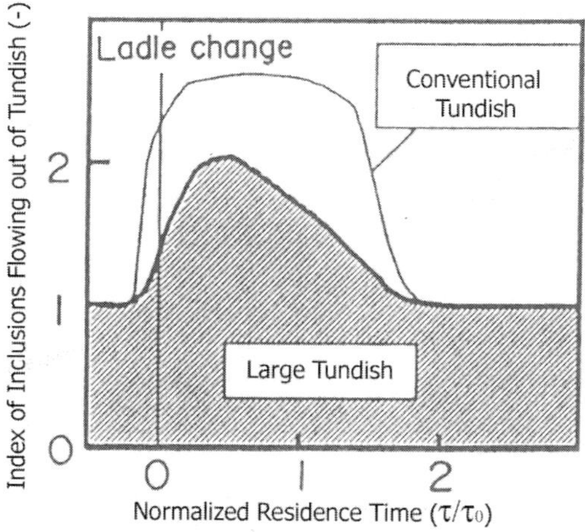

Figure 2.36: Macro inclusions occurring at ladle change reduced by increasing tundish capacity from the conventional size (65-ton) to a large size (85-ton) for casting high end LCAK sheet steel. [Ref. 41]

2.5.5 *Remarks on modeling*

As mentioned earlier, an important source of macro inclusions in actual tundish operation is not limited only to the evolution of large-sized alumina clusters from the small ones. Macro inclusions also generate from the entrainment of slag and scum mostly in the inlet section where ladle slag is carried over during ladle change and scum is created by the reoxidation. Some fraction of macro inclusions also occur in the rest of the tundish length by entraining tundish flux, ladle slag, and scum.

Large-sized alumina clusters also form directly from reoxidation by air leaking from the joints at the slide gate, slide gate/SEN, lid/tundish, and by the oxidizing slag. Such macro inclusions are largely responsible for the process problems during casting and downgrading of the final product. It is important to model and understand the entrainment and formation of macro inclusions but it is more important to deal with the entrainment during the non-steady state when most of the problematic macro inclusions form.

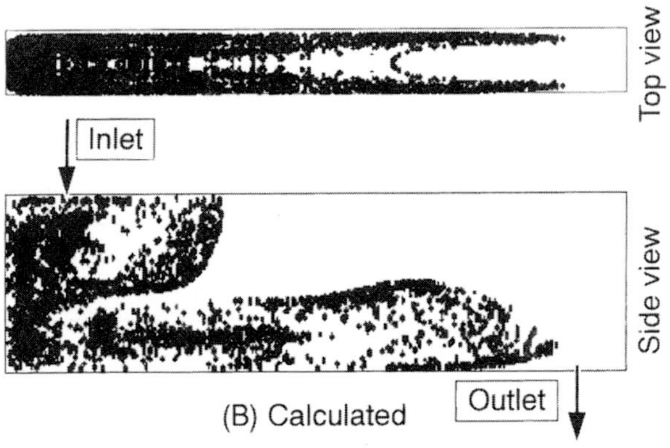

Figure 2.37 (a): Calculated flow pattern of steel melt 30s after inclusion injection in deep tundish with width to depth ratio (W/D) of 0.444. [Ref. 42]

An important function of the tundish is to let the macro inclusions be removed while the melt travels through the tundish. For the time being, modeling has developed to predict the removal of these macro inclusions in the tundish once their size distribution, number density, and site of introduction into tundish are given. More discussions on characterization and modeling of flow in the tundish are given in Chapters 3, 4, and 5. However, the assumptions employed in the previous models require experimental validation to make the model more reliable. Also, the occurrence of the macro inclusions, their size distribution, number density, and site of introduction are quite unpredictable since they depend mostly on site-specific operational conditions. Modeling requires to implement such conditions explicitly and practicably in order to make the model usable for industrial operations. However, most important factor is to prevent the occurrence of the entrainment by improving the tundish hardware system and its operation. These aspects are discussed in Chapter 6.

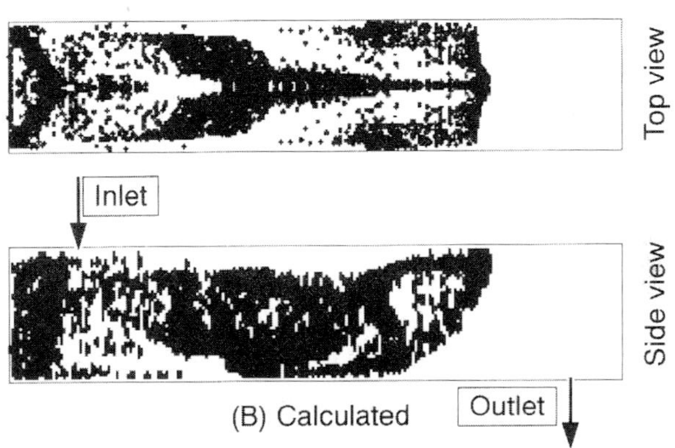

Figure 2.37 (b): Calculated flow pattern of steel melt 30s after inclusion injection in square tundish (*i.e.* W/D = 1.0). [Ref. 42]

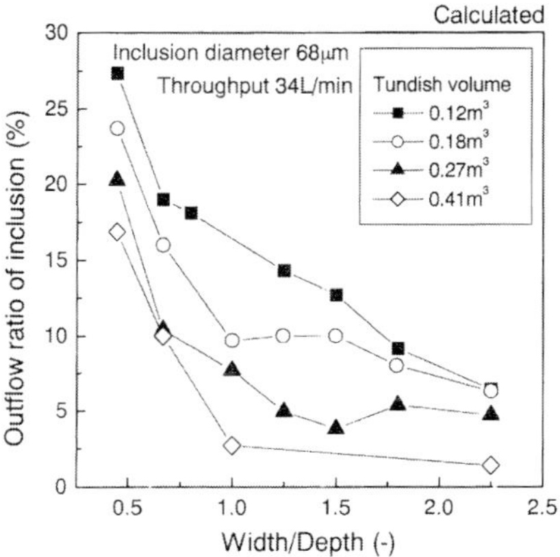

Figure 2.38: Outflow ratio of inclusions as a function of width to depth ratio. [Ref. 42]

In casting plants where (1) tundishes are smaller than ideal, (2) quality demands are not very high, (3) throughput rate of steel melt is limited, and (4) operational skills are less advanced, modeling for steady state casting with tundish furniture is useful. Once clean steel casting during the steady state is successfully established with near ideal tundish design and good operational skills for high quality steels, only issue that remains to be resolved is the occurrence of macro inclusions during the non-steady state casting. Modeling has not been able to take care of this issue yet, and future development of modeling should address this issue.

References

1. T. Emi, *Process integration for making extra clean steels for stringent applications, in "Metal Separation Technologies beyond 2000", eds. K. C. Liddel and D. J. Chaiko,* (TMS, Warrendale, 1999), 207-218.
2. M. Ikeda, in *"History of Continuous Casting Technology of Steel in Japan", (1996, ISIJ, Tokyo),* Chapter 6, 482.
3. Y. Yoshida and Y. Funahashi, *Tetsu-to-Hagane,* 1975, 61, 2489-2500.
4. Y. Habu, H. Kitaoka, Y. Yoshii, T. Emi, Y. Iida and T. Ueda, Tetsu-to-Hagane, 60(1976), pp. 1803-1812.
5. H. Sakao, *"Fundamentals of Steelmaking Reaction-Oxidation Reaction", in Handbook of Iron and Steel 3rd. Ed.,* (1981, ISIJ, Tokyo), vol. 1, Fundamentals, 166 *based on data published in Recommended Equilibrium Values for Steelmaking Reactions, Japan Society for the Promotion of Science, 19th Committee for Steelmaking, (English version, Steelmaking Data Sourcebook, Gordon & Breach Science Publishers, New York,* 1988)
    [cf. J. Chipman and J. F. Elliott, "Electric Steel Making, II", Chapt. 16, (1963, AIME & Interscience. Publishers., N.Y.), 95.
6. C. R. Taylor and J. Chipman: *Trans. Metall. Soc. AIME,* 1943, 154, 228-247.
7. H. Rein and J. Chipman: *Trans. Metall. Soc. AIME,* 1965, 233, 415-425.
8. H. Itoh, M. Hino and S. Banya, Metall. *Mater. Trans. B,* 1997, 28B, 953-956.

9. M. Hino and S. Ban-ya, *14th PTD Conf. Proc.*, (1995, ISS, Warrendale), 47-52.

10. R. Dekkers, B. Blanpain and P. Wollants, *Metall. Mater. Trans. B*, 2003, 34B, 161-171.

11. R. Dekkers, *"Non-Metallic Inclusions in Liquid Steel"*, *Ph. D. Thesis*, Katholieke Universiteit Leuven, June, 2002.

12. Y. Miki, Y. Shimada, B. G. Thomas and A. Denissov, *Iron & Steelmaker*, Aug.1997, 31-38.

13. H. Yin, H. Shibata, T. Emi and M. Suzuki, *ISIJ Internat.*, 1997, 37 936-945, and 945-955.

14. T. Kirihara, K. Tozawa and K. Sorimachi, *CAMP-ISIJ*, 2000, 13, 120.

15. E. Steinmetz, H.-U. Lindenberg et al, *Arch. Eisenhuttenw.*, 1976, 47, 199-204; 1977, 48, 569-574; *Stahl u. Eisen*, 1977, 97, 1154-1159; and 1983, 103, 539-545.

16. T. Malkiewicz and S. Rudnik, *JISI.*, 1963, 201, 33-38.

17. P. J. H. Maunder and J. A. Charles, *JISI*, 1968, 206, 705-715.

18. R. Kiessling and N. Lange, *Non-Metallic Inclusions in Steel, Part-I to –IV, Publ. The Metals Soc., London*, 1978, No.194, R. Kiessling, *ibid., Part-V, Inst. Metals, London,* 1989.

19. P. G. Saffman and J. S. Turner, *J. Fluid Mech.*, 1956,1, 16-30.

20. K. Higashitani, K. Yamauchi, Y. Matsuno and G. Hosokawa, *J. Chem. Eng. Jpn.*, 1983, 16, 299-304.

21. U. Lindborg and K. Torssell, *Trans. Met. Soc. AIME*, 1968, 242, 94-102.

22. S. Taniguchi and A. Kikuchi, *Tetsu-to-Hagane*, 1992, 78, 527-535.

23. S. Linder, *Scand. J. Metall.*,1974, 3, 137-150.

24. K. Nakanishi and J. Szekely, *Trans. ISIJ*, 1975, 15, 522-530.

25. O. Levenspiel, *Chemical Reaction Engineering, John Wiley & Sons, Inc.*, 1972.

26. S. Asai, T. Okamoto, J. C. He and I. Muchi, *Tetsu-to-Hagane*, 1982, 68, 426-434, also *Trans. ISIJ*, 1985, 25, 43-50.

27. K. Mori and M. Sano, *Tetsu-to-Hagane*, 1981, 67, 672-695, also M. Sano and K. Mori, *Trans ISIJ,* 1980, 20, 668-674, 675-681, 1983, 23, 169-175.

28. Y. Sundberg, *Scand. J. Metall.*, 1978, 7, 81-87.

29. D. Mazumdar and R. I. L. Guthrie, *ISIJ Internat.*, 1995, 35, 1-20.

30. T. Sakuraya, N. Sumida, K. Onuma and T. Fujii, *CAMP-ISIJ*, 1988, 1, 225.

31. H. Sandberg, T. Engh, J. Andersson and R. Olsson, *Proc. 1ˢᵗ Japan-Sweden Joint Symp. on Ferrous Metallurgy, ISIJ, TOKYO,* May, 1971, 87-107.

32. T. Fujii Y. Oguchi, N. Sumida, T. Emi and K. Ishizaka, *Tetsu-to-Hagane*, 1982, 68, 1595-1603.

33. Y. Miki, B. G. Thomas, A. Dennisov and Y. Shimada, *Iron and Steelmaker*, 1997, 24, No. 8, 31-39.

34. D.-Y. Sheng, M. Soder, Par Jonsson and L. Jonsson, *Proc. 6th Internat. Conf. on Molten Slags, Fluxes and Salts, Stockholm-Helsinki,* June, 2000 CD-ROM, paper No. 127.

35. T. Sawai, W. Yamada, S. Tanaka, T. Matsumoto, H. Takahashi and R. Takahashi, *CAMP-ISIJ*, 1996, 9, 766.

36. H. Tozawa, Y. Kato, K. Sorimachi and T. Nakanishi, Agglomeration and flotation of alumina clusters in molten steel, *ISIJ Internat.*, 1999, 39, No. 5, 426-434.

37. R. Jullien, M. Kolb and R. Botet, *J. Phys. Lett.*, 1984, 45, L-211.

38. Y. Miki and B. G. Thomas, *Metall. Mater. Trans.B*, 1999, 30B, 639-654.

39. S. Chakraborty and Y. Sahai, *Ironmaking and Steelmaking*, 1992, 19, 488-494.

40. C. Marique, *Proc. 7ᵗʰ Internat. Conf. on Refining Processes, Scanmet-VII,* Lulea, MEFOS, June, 1995, Part II, 107-145.

41. Y. Akai, K. Yamada, M. Toyota, M. Hoteiya, K. Sekino and K. Takatani, *CAMP-ISIJ*, 1991, 4, 1322.

42. T. Nakaoka, T. Miyake, T. Mimura and H. Tai, *Tetsu-to-Hagane*, 2000, 86, 231-238.

Chapter 3

Review of Fluid Flow and Turbulence

3.1 Introduction

Fluid flow plays a very important role in tundish operations, as the laws of fluid mechanics govern the flow of metal, entrainment of gas and slag, and the movement and flotation of non-metallic inclusions. Detailed discussions of fluid flow and turbulence can be found in many textbooks on transport phenomena [e.g. Refs. 1-4]. This chapter presents a brief overview of the subject.

3.2 Fluid Flow Regimes

Fluid flow regimes may be classified in many ways, but one of the most important distinctions is *laminar* and *turbulent* flow regimes. In *laminar flow,* layers of fluid slide over one another without any macroscopic mixing or intermingling of fluid, in the direction perpendicular to the fluid flow. For the flow of a fluid of given physical properties, when the velocity is increased in a conduit of a given geometry there exists a critical velocity above which the flow becomes *turbulent.* In a turbulent flow, macroscopic mixing takes place between fluid layers or portions of fluid over relatively long distances. One of the earliest investigations into turbulent flow was conducted by Osborne Reynolds [5], who established a criterion for the transition from *laminar* to *turbulent* flow by a dimensionless quantity known as the Reynolds number (Re):

$$\mathrm{Re} = \frac{Lv\rho}{\eta} \tag{3.1}$$

where L is the characteristic length of the system (e.g. the diameter of a pipe), v is the average velocity of the fluid, and ρ and η are the density and viscosity of the fluid, respectively. For fluid flow in a smooth, circular pipe this transition takes place at a Re of approximately 2100. The value of the Reynolds number at which the transition from laminar to turbulent flow takes place is strongly dependent on the bounding geometry of the system, owing in part to the arbitrary nature of the characteristic length, L, and in part to inherent differences in the flow pattern. For free surface flow in a channel such as that existing in a tundish, the Re for laminar to turbulent flow is approximately 500, where the characteristic length is the depth of the fluid.

3.3 Newton's Law of Viscosity

Consider a fluid confined between two parallel plates as shown in Fig. 3.1. The upper plate is stationary, and the lower plate is set in motion at a velocity, V, and at time $t = 0$. Assuming that there is no slip between the lower plate and the layer of fluid in contact, this layer of fluid will start moving with the velocity V. Similarly, the layer of fluid in contact with the upper plate will be stationary. The lowermost fluid layer gradually transfers momentum to the upper fluid layers. Thus for some time, an unsteady state flow condition exists, in which velocity at a given point in the system changes with time. After sufficient time has elapsed, at a steady state, a laminar velocity profile is attained, and thus the velocity at a given point in the fluid system does not change with time.

At steady state, a force, F, must be exerted on the lower plate to keep it in motion with the constant velocity, V. Newton found that the force per unit area of the plate is proportional to the velocity, V, and inversely proportional to the spacing, Y, between the plates.

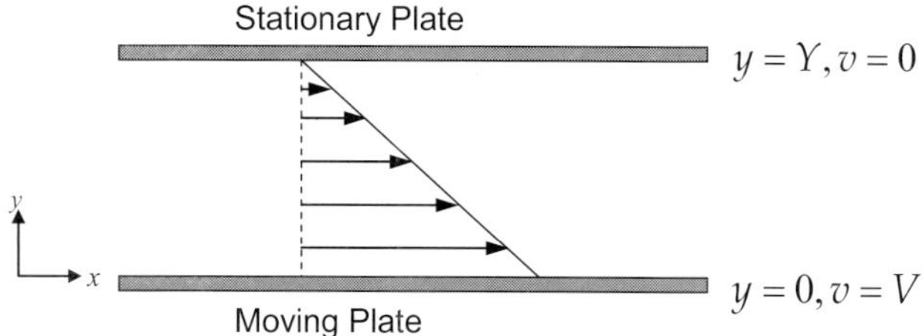

Figure 3.1: Laminar flow of fluid between two plates.

$$\frac{F}{A} \propto \frac{V}{Y} \tag{3.2}$$

where A is the surface area of the plate,

$$\frac{F}{A} = \eta \frac{V}{Y} \tag{3.3}$$

where η is the constant of proportionality.

The force per unit area (F/A) is the shear stress at the interface between the lower plate and the adjacent fluid layer. The velocity gradient in the y-direction in Eq. (3.3) may be expressed in differential form as dv_x/dy, and the shear stress as τ_{yx}. Thus Eq. (3.3) may be written as

$$\tau_{yx} = -\eta \frac{dv_x}{dy} \tag{3.4}$$

The first subscript y of τ_{yx} refers to the direction of the momentum transfer, and the second subscript, x, refers to the direction of fluid velocity. η is known as the *coefficient of molecular viscosity* or simply the *viscosity* of the fluid. In this case, the velocity gradient, dv_x/dy, is negative, and viscosity is always positive. Thus, the minus sign in Eq. (3.4) makes the shear stress, τ_{yx}, positive. Shear stress, τ_{yx}, is the transport of momentum in the y-direction per unit time per unit area, or the rate of momentum transport per unit area. In the flow shown in Fig. 3.1, the momentum is being transferred in the positive y-direction.

Eq. (3.4) is *Newton's law of viscosity*. According to Eq. (3.4), a plot of shear stress, τ_{yx}, against the velocity gradient, dv_x/dy, should be a straight line passing through the origin, and the slope of the line is the viscosity of the fluid. Thus, in such systems if shear stress is increased, the velocity gradient increases in the same proportion. All fluids which follow this relationship or Newton's law of viscosity are known as *Newtonian fluids*. On the other hand, there is another class of fluids which do not obey this law, such as molten plastics, molten glass, and slurries, etc., and these are termed *non-Newtonian fluids*.

3.3.1 *Viscosity*

In SI units, τ_{yx} has the units of $N/m^2 = kg/m \cdot s^2$. The velocity gradient, dv_x/dy, uses the units of $(m/s)(1/m) = 1/s$. Therefore viscosity, η, has the units of $kg/m \cdot s$, and may also be written as

$$\frac{kg}{m.s} = \frac{N}{m^2} s = Pa \cdot s$$

In the cgs system, the units of viscosity are in $g/cm \cdot s$ which is known as a *poise*:

$$1 \text{ poise } or \ 1P = 1 \ g/cm \cdot s$$

$$1P = 0.1 \ Pa \cdot s$$

Sometimes a more convenient unit for liquids and gases is the *centipoise* (cP), where $1 \ cP = 10^{-2} \ P$.

The viscosity of water at room temperature is approximately $1\,\text{cP}$ or $10^{-3}\,\text{Pa}\cdot\text{s}$.

In some engineering problems, it is convenient to define the ratio of viscosity to density of fluid as the kinematic viscosity, ν. Thus

$$\nu \equiv \frac{\eta}{\rho}$$

The kinematic viscosity is a measure of *momentum diffusivity* analogous to thermal and mass diffusivities. The units of kinematic viscosity are m^2/s, or in the cgs system units are cm^2/s, which is called a *stoke*. As with the viscosity units in the cgs system, the *centistoke*, cS, (0.01 stoke) is also commonly used.

3.4 Dimensionality of Flow

It is very common to refer to any flow as one-, two-, or three-dimensional. Consider the flow system between two parallel plates shown in Fig. 3.1. As described above, flow is in the x-direction, and viscous momentum is being transferred in the y-direction. The third (z-direction) is perpendicular to the x- and y-directions and thus, crosses through the plane of the paper. In any system if velocity changes in any one coordinate direction only, it is considered *one-dimensional flow*. Thus in Fig. 3.1 for given values of x and z, velocity changes only in the y-direction, or velocity is a function of the y-coordinate. Thus, this is a one-dimensional flow.

Consider fluid flowing with a constant stream velocity of v_∞, which is independent of x, y, z, and time. Now a horizontal flat plate is brought into the flow as shown in Fig. 3.2. After sufficient time, a steady state velocity profile is attained in the boundary layer region. It can be seen that near the plate, for given values of x and z, velocity changes in the y-direction. Also, for given values of y and z, velocity changes in the x-direction. Thus in this boundary layer flow, velocity is a function of the *x*- and *y*-coordinates, and thus the flow is two-dimensional. Detailed description of boundary layer theory is provided in Ref. [6].

In general, the flow in any common tundish depends on all three coordinate directions, and thus is a three-dimensional flow. It may be a three-dimensional steady state or unsteady state (transient) flow.

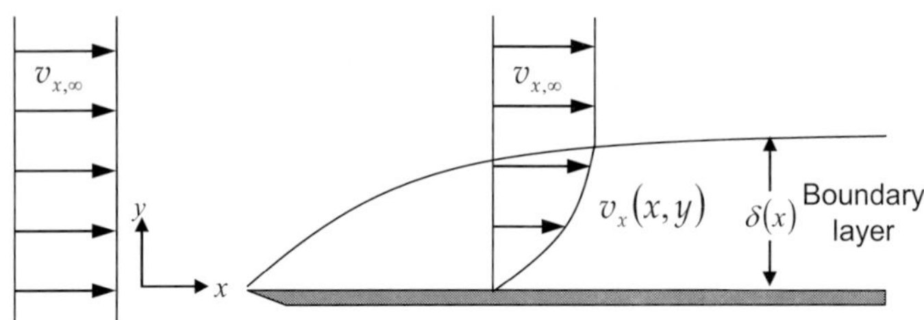

Figure 3.2: Momentum boundary layer on a horizontal flat plate immersed in a flowing fluid.

3.5 Modes of Momentum Transport

Referring to Fig. 3.1, it has been stated above that the momentum transport is in the positive *y*-direction from higher velocity layers to lower velocity layers. This momentum transport is known as *viscous or diffusive* momentum transport. The other momentum transport is due to the motion of the fluid itself in the flow direction, which is termed *convective* momentum transport.

3.5.1 *Viscous or diffusive momentum transport*

The mechanism of this momentum transport may be understood by considering a plane, or the layer of fluid at $y = y_1$ in Fig. 3.1. This and two other layers, one on either side of this plane are schematically shown in Fig. 3.3. The molecules have an average velocity of v_x in the *x*-direction, but in the *y*-direction their average velocity component is zero.

Hence, molecules cross the plane in both directions with equal frequency, but those crossing from a faster moving layer to a slower moving layer possess more x-direction momentum than those crossing from a slower to a faster moving layer. As a result, there is a net transfer of x-direction momentum in the upward or positive y-direction. It can also be shown that τ_{yx} is the rate of momentum transport per unit area.

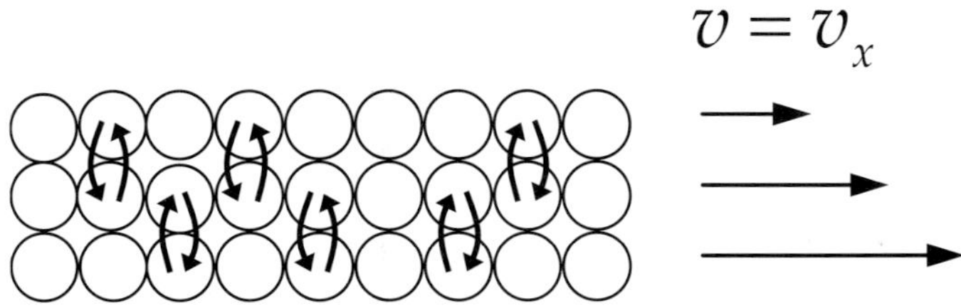

Figure 3.3: Molecules crossing layers and transferring viscous momentum.

3.5.2 *Convective momentum transport*

Convective momentum transport results from the motion of a fluid and is in the direction of the flow. Thus, the rate of convective transport is (mass x velocity)/time, and is equal to $\rho v_x^2 A$, where A is the cross-sectional area perpendicular to the flow direction through which the fluid of density ρ flows. The directions of the viscous and convective momentum transports are shown in Fig. 3.4.

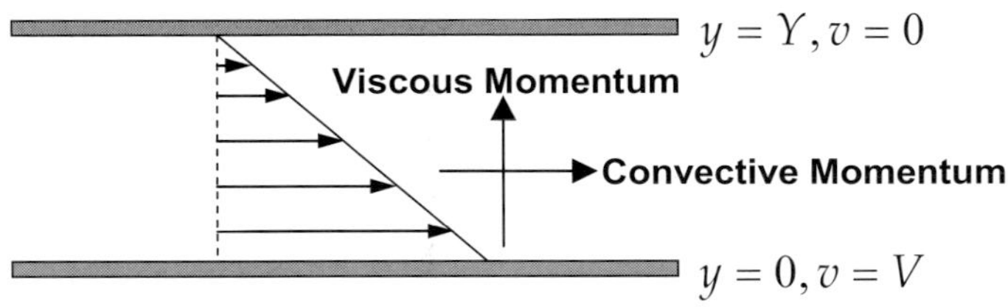

Figure 3.4: Directions of viscous and convective momentum.

3.6 Equations of Continuity and Motion

The equations of continuity and motion are presented in this section, but their derivation and detailed discussion are not presented here. Such discussion is available in any standard textbook on transport phenomena [e.g. Ref. 1]. The equation of continuity is developed by writing a mass balance over a stationary volume element through which fluid is flowing. This equation in vector notation is

$$\frac{\partial \rho}{\partial t} = -(\nabla \cdot \rho v) \tag{3.5}$$

A special form of the equation of continuity for fluid of constant density is as follows:

$$(\nabla \cdot v) = 0 \tag{3.6}$$

Similarly, the equation of motion may be obtained by writing the momentum balance over a volume element. Thus, the equation of motion in single vector notation is

$$\frac{\partial}{\partial t} \rho v = -[\nabla \cdot \rho v v] - \nabla p - [\nabla \cdot \tau] + F_b \tag{3.7}$$

The term on the left side of Eq. (3.7) is the rate of change of momentum per unit volume. The four terms on the right side are the rate of momentum gain by convection per unit volume, pressure force per unit volume, rate of momentum gain by viscous transfer per unit volume, and body force per unit volume, respectively. Body forces may be produced by gravity, natural convection (arising from temperature gradients), or by electromagnetic forces, etc.

For a fluid of constant ρ and η, Eq. (3.7) may be simplified with the use of the equation of continuity (Eq. (3.6)) to give

$$\rho \frac{Dv}{Dt} = -\nabla p + \eta \nabla^2 v + F_b \tag{3.8}$$

where

$$\frac{D(\ )}{Dt} \equiv \frac{\partial(\ )}{\partial t} + v \cdot \nabla(\ ) \tag{3.9}$$

Eq. (3.8) is the famous *Navier-Stokes equation*.

3.7 Stokes' Law

Inclusions are lighter than molten steel and rise up to the free surface where they may get absorbed by the tundish slag. In a quiescent melt, the terminal rising velocity, V_t of an inclusion particle is given by the following relationship, known as *Stokes' law*.

$$V_t = \frac{2R_p^2 (\rho_s - \rho_p) g}{9\eta_s} \tag{3.10}$$

where R_p is the particle radius, ρ_s and η_s are the density and viscosity of molten steel, respectively, ρ_p is the particle density, and g is acceleration due to gravity. Strictly speaking, Stokes' law is valid for particles with particle Reynolds number, Re_p, of less than unity.

$$\mathrm{Re}_p = \frac{D_p V_t \rho_s}{\eta_s} \tag{3.11}$$

where D_p is the particle diameter. In flowing melt such as that in a tundish, melt velocity has to be vectorially added to the rising velocity calculated by the Stokes' law.

3.8 Turbulent Flow

It has been stated above in section 3.1 that the flow transition from laminar to turbulent regimes occurs at a critical Reynolds number. The Re expressed by Eq. (3.1) is the ratio of the inertial to viscous forces. The character of any flow depends upon the relative magnitudes of these two forces acting on the fluid. Inertial forces tend to disrupt the orderly and unidirectional flow, whereas viscous forces tend to stabilize the flow and keep the flow streamlines together. At a relatively low velocity and thus at a low Re, viscous forces dominate and their stabilizing effect makes the flow laminar. However, as velocity increases, the effect of inertial forces dominates, and above a critical Re the flow becomes turbulent. The inertial force acting on an element of fluid is $\rho L^2 v^2$, and the viscous force is $\eta v L$.

$$\text{Re} = \frac{\text{Inertial Force}}{\text{Viscous Force}} = \frac{\rho L^2 v^2}{\eta v L} = \frac{L v \rho}{\eta}$$

which is the definition of Re given by Eq. (3.1).

In a fully developed steady state turbulent flow, velocity at any location fluctuates about its time-averaged value. Thus, the instantaneous velocity of fluid, v_x is a sum of its time-averaged value, $\tilde{v}_x$ and its fluctuating component, v'_x. The instantaneous velocity is therefore

$$v_x = \tilde{v}_x + v'_x \tag{3.12}$$

where time-averaged velocity is defined as

$$\tilde{v}_x = \frac{1}{t} \int_0^t v_x dt \tag{3.13}$$

Fig. 3.5 shows the instantaneous velocity in water measured by a non-intrusive probe. The time-averaged velocity is 370 mm·s^{-1}; however, instantaneous velocity values fluctuate between approximately 290 and 475 mm·s^{-1}.

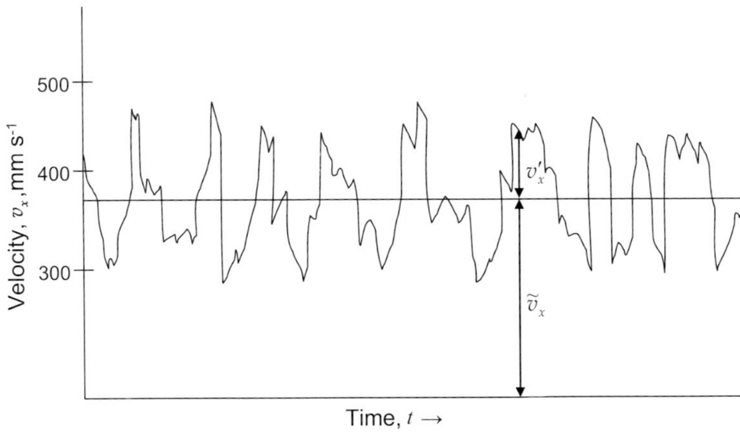

Figure 3.5: Instantaneous velocity measured by Laser Doppler Anemometer.

Thus the time-averaged value of the fluctuating component (or $\overline{v'_x}$) will be zero. But the time-averaged value of the square of the fluctuating component, $\overline{v'^2_x}$ will not be zero, and in fact, $\overline{v'^2_x}\big/\widetilde{v}_x$ is a measure of turbulence level known as *intensity of turbulence.* In a unidirectional (in *x*-direction) turbulent flow, velocity fluctuations in the *y*- and *z*-directions may exist, but time-averaged values of v'_y and v'_z will also be zero. When the fluctuating components in three directions are approximately the same or direction independent, such turbulence is known as *isotropic.* Thus, in isotropic turbulence

$$v'_x = v'_y = v'_z \qquad (3.14)$$

3.8.1 *The eddy size spectrum*

The random velocity fluctuations are consequences of the swirl or eddies generated by a turbulent flow and are inherent characteristics of such a flow. Figures 3.6 and 3.7 show eddies generated in a water flow in an open channel and their schematic representation, respectively. It is evident from Figs. 3.6 and 3.7 that eddies of different sizes from relatively large to very small exist in any fully developed turbulent flow.

Fig. 3.8 shows kinetic energy of turbulence against the wave number or inverse of eddy length plotted in schematic form. The figure shows that the formation of the largest eddies of permanent character are dependent on the flow conditions. They contain a smaller amount of turbulent kinetic energy, which they derive from the main flow. The interactions between these large eddies result in the formation of medium sized eddies which contain relatively large energy, and thus are termed *energy containing eddies*. The average size of the energy containing eddies is ℓ_e. They feed energy to the smaller eddies, which in turn lose the energy in the form of heat through viscous dissipation. This range of smaller eddies is known as the *"universal equilibrium"* range.

Figure 3.6: Eddies in water flow in a channel. [Ref. 7]

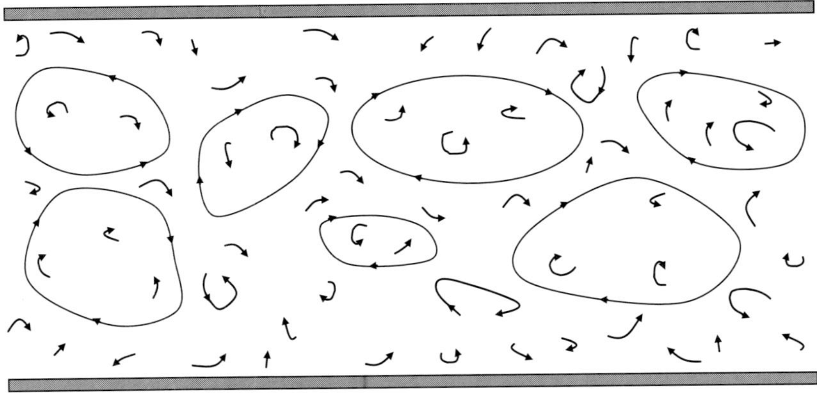

Figure 3.7: Schematic representation of distribution of eddies.

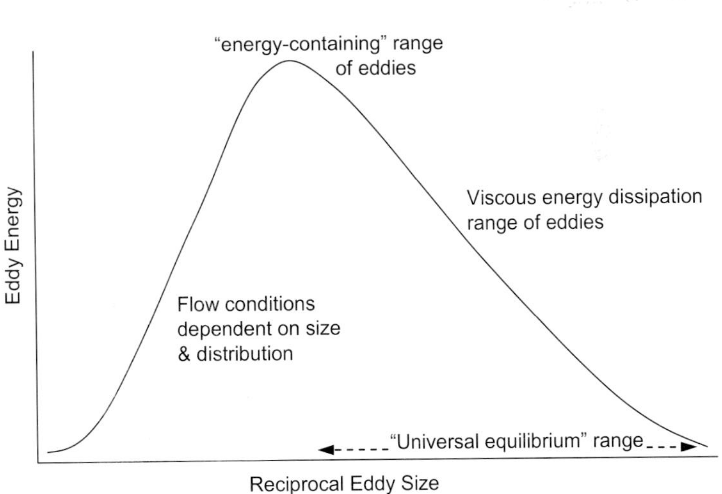

Figure 3.8: Eddy spectrum and its energy.

3.8.2 *Prandtl's mixing length and effective viscosity*

Prandtl considered an analogy between the transfer of viscous momentum due to exchange of molecules in adjacent layers in laminar flow, and turbulent momentum transfer due to movement of eddies over relatively large distances. This is known as *mixing length* or *eddy length*. Thus, analogous to Newton's law of viscosity, Prandtl proposed

$$\tau_{yx,t} = -\eta_t \frac{d\tilde{v}_x}{dy} \qquad (3.15)$$

where $\tau_{yx,t}$ is the turbulent shear stress, and η_t is the turbulent viscosity generated as a result of movement of eddies over relatively large distances (eddy length) compared to the movement of molecules over the mean free path in laminar flow. Thus, the values of turbulent viscosity may even be 10^3 to 10^5 times greater than the molecular viscosity of a fluid. The term *effective viscosity*, η_{eff} is used for the sum of molecular viscosity, η and turbulent viscosity, η_t. In a turbulent flow, η is negligible in comparison to η_t, and hence the turbulent and effective viscosity values are almost the same.

3.8.3 *Turbulent shear stress*

It is well known that the shear stress (i.e. resistance to flow) is much higher in a turbulent flow compared to a laminar flow. The reason is the continuous exchange of packages of fluid which leave one region and move to a different region traveling at a different velocity. Thus, it causes either a momentum gain or loss and results in higher shear stress.

Consider turbulent flow in a pipe, as shown in Fig. 3.9, with lower average velocity near the wall and faster flow away from the wall. If a packet of fluid moves from a slowly moving layer to faster moving layer, it reduces the x-direction momentum at the new location. At any instant if the fluctuating component of velocity in the y-direction is v'_y, the momentum transfer in the positive y-direction will be $\rho v'_y$ per unit volume of fluid. Since this packet is coming from a slower moving layer and the momentum is conserved, it will cause an instantaneous reduction in momentum, causing a fluctuating component of v'_x in the negative direction. Thus, the rate of momentum change per unit area due to this transfer is given by $\rho v'_y v'_x$. As explained above, the signs of v'_y and v'_x will always be the opposite, and thus $-\rho v'_y v'_x$ will be positive. This is equal to the turbulent shear stress, *i.e.* the rate of change of momentum per unit area. Thus at any instant of time,

$$\tau_{yx,t} = -\rho v'_y v'_x \tag{3.16}$$

Time-averaged turbulent shear stress will be $-\rho \overline{v'_y v'_x}$ which is commonly termed the *Reynolds stress*. Combining this with Eq. (3.15), we get

$$\tau_{yx,t} = -\eta_t \frac{d\tilde{v}_x}{dy} = -\rho \overline{v'_y v'_x} \tag{3.17}$$

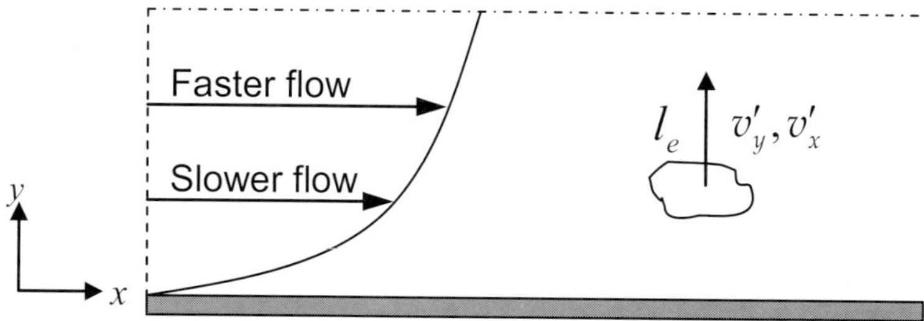

Figure 3.9: Turbulent flow in a pipe and associated shear stress.

3.9 Turbulent Equations of Continuity and Motion

Turbulent equations of continuity and motion may be obtained by replacing the velocities in laminar flow equations (Eqs. (3.6) and (3.7)) by $\widetilde{v}_x + v'_x$. Thus the following time-averaged equations will be obtained

$$\left(\nabla \cdot \widetilde{v}\right) = 0 \tag{3.18}$$

$$\rho \frac{D\widetilde{v}}{Dt} = -\nabla p + \eta_{eff} \nabla^2 \widetilde{v} + F_b \tag{3.19}$$

It can be seen that these equations are similar in mathematical formulation to the laminar flow equations. The molecular viscosity has been replaced by the effective viscosity, and the velocities have been replaced by time-averaged velocities.

3.10 Heat and Mass Transfer

Heat and mass flux under turbulent conditions can be expressed by the following equations, which are analogous to the turbulent shear stress given by Eq. (3.15);

$$q = -k_{eff} \frac{dT}{dy} \tag{3.20}$$

$$J = -D_{eff} \frac{dC}{dy} \tag{3.21}$$

where q and J are the rates of heat and mass flux, k_{eff} and D_{eff} are effective transport coefficients, that is, the effective thermal conductivity and diffusivity which are the sum of molecular and turbulent components, and dT/dy and dC/dy are the temperature and concentration gradients, respectively.

Turbulent transport equations for heat and mass transfer can be mathematically represented by the following equations similar to the turbulent momentum transport equation (Eq. (3.19)).

Heat Transfer
$$\frac{DT}{Dt} = \alpha_{eff} \nabla^2 T + S_T \tag{3.22}$$

Mass Transfer
$$\frac{DC}{Dt} = D_{eff} \nabla^2 C + S_c \tag{3.23}$$

where T and C are temperature and concentration, α_{eff} and D_{eff} are effective thermal and mass diffusivities, and S_T and S_c are the source terms for the corresponding variables, respectively. All terms in Eqs. (3.22) and (3.23) are similar to Eq. (3.19) with the difference that the pressure term is missing in the heat and mass transport equations. The effective transport coefficients in Eqs. (3.19), (3.22), and (3.23) are the sum of the corresponding molecular and turbulent properties. Since in a turbulent flow system, the molecular component of these transport coefficients is negligible in comparison to its turbulent counterpart, the effective coefficient is practically the same as its turbulent component. Also, the transport takes place due to the movement of fluid packets over equivalent Prandtl mixing length distances, and thus it is safe to assume that the turbulent Prandtl (v_t/α_t) and turbulent Schmidt (v_t/D_t) numbers are equal to unity, where v_t is turbulent kinematic viscosity. The turbulent transport coefficient has to be deduced from an additional relationship usually known as a *turbulence model*, which sometimes may be quite complex.

3.11 Turbulence Models

Many models are presently available that have been used for flow field calculations. A good description of many of these models and their relative advantages and disadvantages is given by Launder and Spalding [8]. The simplest turbulence models provide algebraic equations for

     Tundish Technology for Clean Steel Production

calculation of turbulent viscosity in a calculation domain. The turbulent viscosity may be a function of location in the flow field or may even be constant (time and space independent). Such models may be very useful for the solution of inertially dominated flows. Sometimes they are employed to obtain a preliminary or rough solution of the flow system. These models are also very useful during the development and testing of a new model in its early stages. Use of such turbulence models is also less costly in terms of computer time.

More sophisticated models involve the solution of differential equations. These models calculate the intensity and length scale of turbulence as a function of location in the flow field. Many of these models have been developed, but the one that has been the most extensively used is the so-called "K-ε" two-equation model of Launder and Spalding [9]. In this equation, K is kinetic energy of turbulence per unit mass, and ε is rate of energy dissipation. Thus,

$$K = \frac{1}{2}\left(\overline{v_x'^2} + \overline{v_y'^2} + \overline{v_z'^2}\right) \tag{3.24}$$

or in isotropic turbulence,

$$K = \frac{3}{2}\left(\overline{v_x'^2}\right) \tag{3.25}$$

K and ε are estimated from the following differential equations, which represent transport of K and ε in the same way as transport of heat and mass is given by Eqs. (3.22) and (3.23).

$$\frac{DK}{Dt} = \frac{v_t}{\sigma_k}\nabla^2 K + G_k - \varepsilon \tag{3.26}$$

$$\frac{D\varepsilon}{dt} = \frac{v_t}{\sigma_\varepsilon}\nabla^2\varepsilon + \frac{\varepsilon}{K}\left(C_1 G_K - C_2\varepsilon\right) \tag{3.27}$$

where G_K is rate of production of K and is given by the following equation:

$$G_k = v_t \left(\frac{\partial v_i}{\partial x_j} + \frac{\partial v_j}{\partial x_i} \right) \frac{\partial v_i}{\partial x_j} \qquad (3.28)$$

The turbulent viscosity is calculated by the following equations:

$$\eta_t = \frac{C_\eta \rho K^2}{\varepsilon} \qquad (3.29)$$

and
$$\eta_{eff} = \eta + \eta_t \qquad (3.30)$$

C_1, C_2, C_η, σ_K, and σ_ε are constants of the model, and their values are suggested by Launder and Spalding [9].

3.12 Concluding Remarks

This chapter has provided a very brief review of some of the fundamentals of fluid flow which are used in the subsequent chapters. These fundamentals form the basis of studying any design of an existing tundish or in designing a new tundish with optimal flow characteristics.

References and Further Reading

1. R. B. Bird, W. E. Stewart, and E. N. Lightfoot, *Transport Phenomena,* Wiley, New York, 1960.
2. J. T. Davies, *Turbulence Phenomena,* Academic Press, New York, 1972.
3. H. Tennekes, and J.L. Lumley, *A First Course in Turbulence,* The MIT Press, 1972.
4. J. O. Hinze, *Turbulence,* McGraw-Hill, New York, 1975.
5. O. Reynolds, *Phil. Trans. Roy. Soc. London,* 1883, 174A, 3, 935.
6. H. Schlichting, *Boundary Layer Theory,* McGraw-Hill, New York, 1979.

7. J. Szekely, *Fluid Flow Phenomena in Metals Processing,* Academic Press, London, 1979.

8. B. E. Launder, and D. B. Spalding, *Lectures in Mathematical Models of Turbulence,* Academic Press, New York, 1972.

9. B. E. Launder, and D. B. Spalding, *Comput. Meth. Appl. Mech. Engng.,* 1974, 3, 269.

Chapter 4

Fluid Flow Characterization

4.1 Introduction

The tundish in a continuous casting operation is an important link between the ladle, a batch vessel, and the casting mold with a continuous flow of metal. The tundish has traditionally served as a metal reservoir and as a distributor. However, in the last couple of decades or so, the role of the tundish has expanded considerably. Because this is the last vessel before solidification of the metal in the mold, tundish operation should ensure that liquid metal of desired temperature, cleanliness, and composition is delivered at a desired volumetric flow rate into the mold. Thus, inclusion flotation and separation, along with composition adjustment have become important functions of the tundish. The efficiency and optimization of these processes require close control of the molten steel flow characteristics within the tundish. If the flow of metal in the tundish is not properly controlled, the quality of steel produced in the ladle may even deteriorate. This chapter deals with the fundamentals of fluid flow characterization pertinent to continuous casting tundish systems. More detailed description of this topic is provided by Levenspiel [1]. Chapter 5 will discuss the application of these principles for the modeling of molten steel flow in tundishes.

4.2 Stimulus-Response Techniques

A detailed characterization of metal flow in a tundish or in any flow system requires knowledge of the complete fluid flow pattern. Because of the practical difficulties connected with obtaining and interpreting such information, an alternate approach is generally used. This approach requires only the knowledge of how long different elements of the fluid remain in the vessel. This partial information is relatively simple to obtain experimentally, can be easily interpreted, and yields information which is sufficient in many cases to allow a satisfactory accounting of the actual existing flow pattern. It produces success either with or without the use of flow models.

The experimental technique used for finding this desired distribution of residence times of fluid in the vessel is known as the "stimulus-response technique." The stimulus or input is simply an addition of a tracer material (e.g. dye, salt, acid, radioactive material, metal solute, etc.) into the fluid stream entering the vessel. The way the tracer is added or the input signal may be of any type: a random signal, a cyclic signal, a step or jump signal, a pulse or discontinuous signal. The response or output signal is then the detection of the tracer leaving the vessel. Although any type of tracer input signal may be used in such studies, the step and pulse input methods are more common for tracer injection. The response is plotted as a dimensionless concentration-time curve which represents the Residence Time Distribution (RTD) of the fluid.

If V is the volume of fluid in a vessel flowing at a volumetric flow rate of Q, then, for an incompressible fluid, the average time spent by the fluid in the reactor is given by

$$\bar{t} = \frac{V}{Q} \tag{4.1}$$

where $\bar{t}$ is known as the *theoretical average residence time* or the *nominal holding time* of the fluid in the vessel.

The RTD of the fluid in a vessel is plotted as a dimensionless time against a dimensionless concentration. The dimensionless time, θ, which is an indication of the fractional residence, is obtained by dividing any time by the theoretical average residence time. Thus,

$$\theta = \frac{t}{\bar{t}} \tag{4.2}$$

The dimensionless concentration for the step input of tracer, F, is given by

$$F = \frac{c}{c_i} \tag{4.3}$$

where c is any concentration of tracer in the fluid at the exit of the vessel, and c_i is the tracer concentration in the incoming fluid.

The dimensionless concentration for the pulse input of tracer, C, is given by

$$C = \frac{c}{q/V} \tag{4.4}$$

where q is quantity of tracer, and q/V is the average concentration of the tracer when it dissolves in the fluid volume, V, in the vessel.

4.2.1 *Step input and F-curve*

Consider a flowing system, with no tracer initially. At a given time (say at $t = 0$), a step change of tracer concentration is made to the fluid stream. A concentration-time curve obtained at the outlet stream plotted on a dimensionless scale is termed the *F-curve*. Figure 4.1 shows a step input and an F-curve which rises from a concentration of 0 to 1. Due to the continuous injection of tracer, the outlet stream tracer concentration attains a value equal to the inlet stream concentration. The use of the step tracer input may not be practical for a number of reasons, including extensive contamination of the product and the cost of tracer.

4.2.2 *Pulse input and C-curve*

When an injection of tracer is made as a short pulse, the resulting dimensionless concentration-time curve at the outlet stream plotted on a dimensionless scale is termed the *C-curve*. Figure 4.2 shows a pulse input at time, t or $\theta = 0$, and the resulting C-curve. Such an input is also called a delta function. The output concentration rises to some value and drops again to zero when the entire tracer has exited the vessel. Plotted on a dimensionless scale, the area under the curve is always unity. Thus

$$\int_0^\infty C d\theta = 1 \tag{4.5}$$

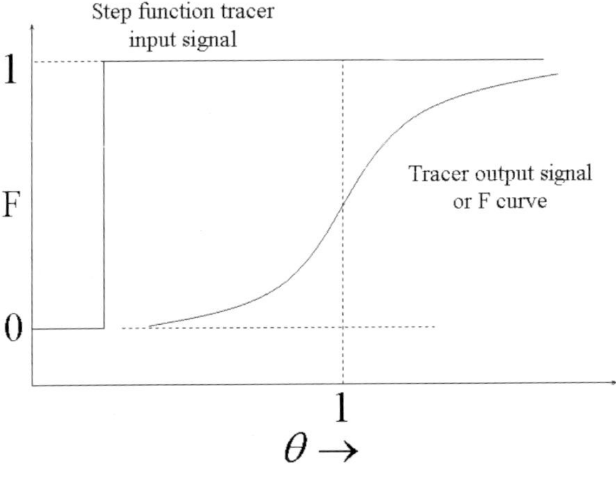

Figure 4.1: A step tracer input and a typical response signal, called the *F-curve*.

4.3 Characterization of Flow Systems

Continuous flow systems may be classified by the type of flow they exhibit. The two extreme or limiting cases of possible fluid flow, termed *plug flow* and *well-mixed flow*, may be considered *ideal* flows. In actuality, the flow systems deviate considerably from these limiting cases; such flows are called *non-ideal* flows.

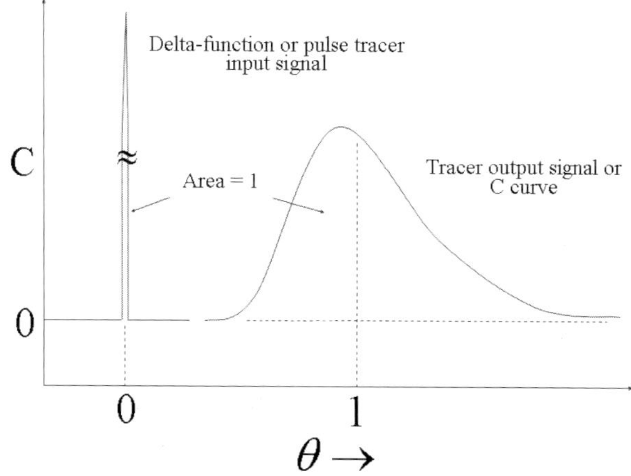

Figure 4.2: A pulse tracer input and a typical response signal, called the *C-curve*.

4.3.1 *Plug flow*

In this limiting type of ideal flow system, longitudinal mixing is non-existent. However, there may be transverse mixing to some extent in the vessel. The necessary and sufficient condition for a plug flow system is that all fluid elements have an identical residence time (equal to the mean residence time) in the vessel. This type of flow is also called "piston flow." Plug flow is schematically represented in Fig. 4.3.

The C-curve response or output of a pulse input in a plug flow system is shown in Fig. 4.4. In the absence of longitudinal mixing in a plug flow reactor, the input pulse is simply shifted by the mean residence time. In other words, the injected tracer pulse exits after the dimensionless time, θ, of 1. Similarly, the F-curve in response to a step input has an identical shape to the input signal, except that it is also shifted by the mean residence time. As shown in Fig. 4.5, the original fluid is completely replaced by the tracer fluid after the mean residence time.

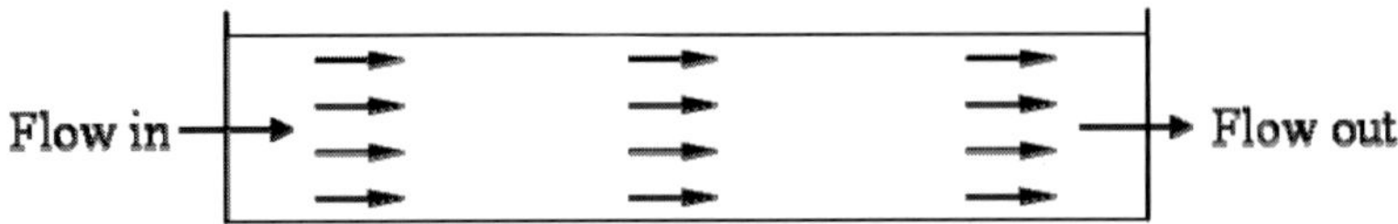

Figure 4.3: Schematic representation of the plug flow.

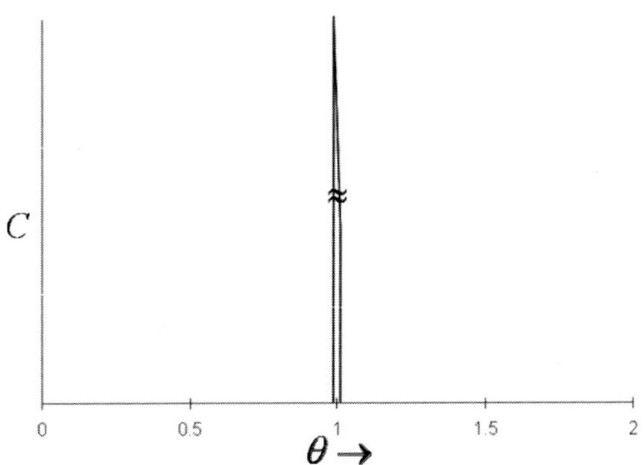

Figure 4.4: C-curve for the plug flow system.

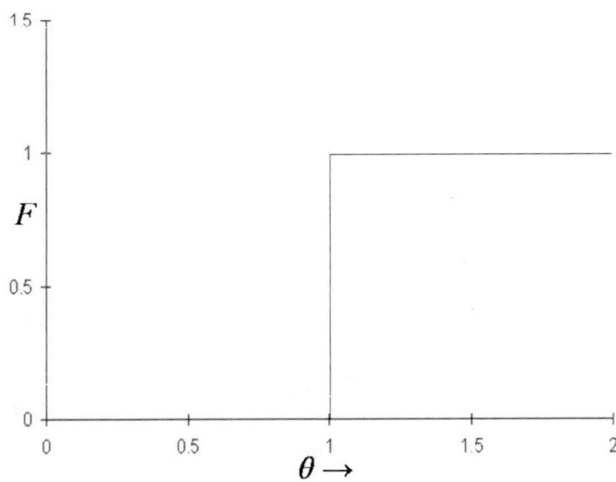

Figure 4.5: F-curve for the plug flow system.

4.3.2 *Well-mixed flow*

This type of ideal flow represents the other extreme of mixing, i.e. the maximum possible mixing of the fluid in the vessel. As a consequence, the outlet concentration and the bulk concentration in the vessel remain the same. A continuous flow system with well-mixed flow is schematically shown in Fig. 4.6.

Consider a mass balance of tracer in the system:

Rate of tracer input – Rate of tracer output = Rate of tracer accumulation

$$Q(0) - Qc = \frac{d}{dt} Vc \tag{4.6}$$

$$\frac{dc}{c} = -\frac{Q}{V} dt \tag{4.7}$$

Upon integration with the boundary conditions such that $c = q/V$ at $t = 0$ and $c = c$ at $t = t$,

$$\ln \frac{c}{q/V} = \ln C = -\theta \tag{4.8}$$

$$C = e^{-\theta} \tag{4.9}$$

The RTD curve given by Eq. (4.9) is shown in Fig. 4.7.

Fig. 4.7 shows the C-curve for a well-mixed flow after a tracer quantity, q, has been dissolved in the entire volume of fluid in the vessel. The residence time distribution of the fluid follows an exponential decay, with the exit stream composition being identical to that within the mixing tank. By the definition of well-mixed flow, the first fluid exiting the vessel will have an average concentration of q/V. Thus, the C-curve starts at a dimensionless concentration of unity. The concentration at the outlet stream gradually decreases as the amount of tracer in the vessel decreases.

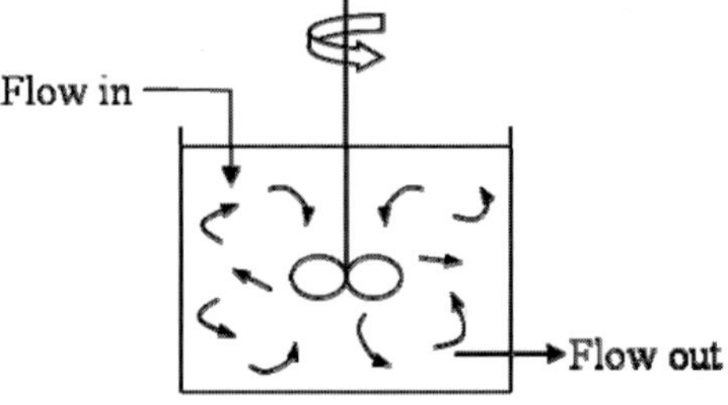

Figure 4.6: Schematic representation of well-mixed flow.

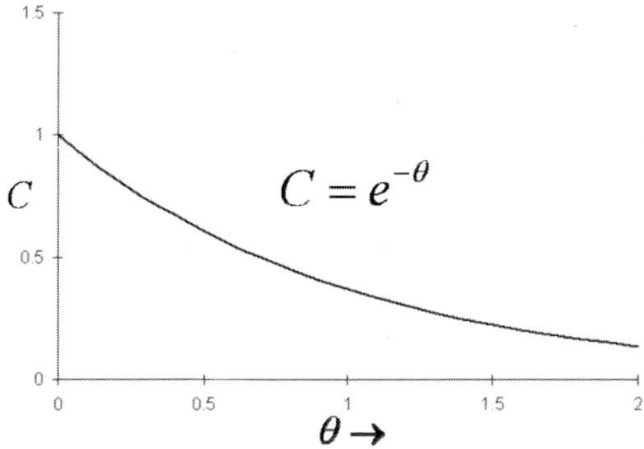

Figure 4.7: C-curve for a well-mixed flow system.

When a step change in the input is imposed on the incoming stream of the vessel with well-mixed flow, the F-curve output increases gradually from 0 to 1, as shown in Fig. 4.8.

Consider the mass balance for a tracer in the system:

Rate of tracer input – Rate of tracer output = Rate of tracer accumulation

$$Qc_i - Qc = \frac{d}{dt}Vc \qquad (4.10)$$

$$d\theta = \frac{dF}{1-F} \qquad (4.11)$$

Upon integration with the boundary conditions such that $F = 0$ at $\theta = 0$ and $F = F$ at $\theta = \theta$, one gets

$$F = 1 - e^{-\theta} \qquad (4.12)$$

The RTD curve given by Eq. (4.12) is shown in Fig. 4.8.

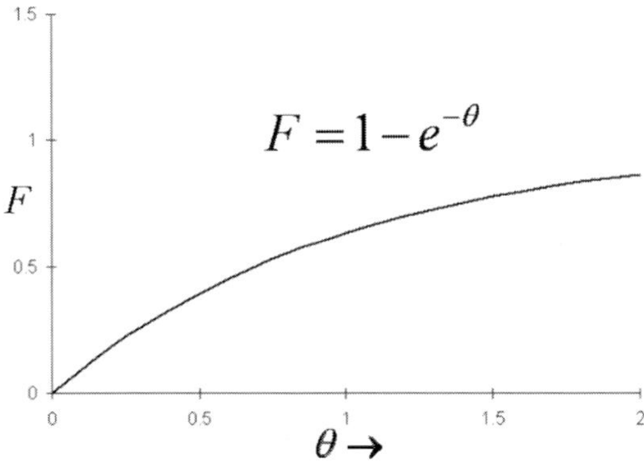

Figure 4.8: F-curve for the well-mixed flow system.

4.4 Characterization of Actual Systems

As stated previously, longitudinal mixing is non-existent in a plug flow system, and at the other extreme, mixing is at a maximum in the well-mixed system. All actual flow systems lie between these two limits. This section describes models which have been used for quantitative flow characterization in actual systems. These models contain certain parameters which can be adjusted so that the predictions of the model match the experimental data. The values of the best fitting parameters most accurately describe the mixing in the vessel.

4.4.1 *Longitudinal dispersion model*

Figure 4.9 shows tracer locations at different times, injected into a plug flow at time, $t = 0$. Being a plug flow, the tracer flows through the system without any longitudinal mixing and travels some distance in times t_1 and t_2. The longitudinal dispersion model or dispersed plug flow model assumes that some extent of turbulent eddy dispersion occurs in the plug flow in this system. Figure 4.10 shows the superimposition of longitudinal dispersion on the plug flow. As a result, the tracer injected as a pulse spreads over some distance at time t_1, and the spread of tracer increases at time t_2. A similar dispersion following a step input of tracer is shown in Fig. 4.11.

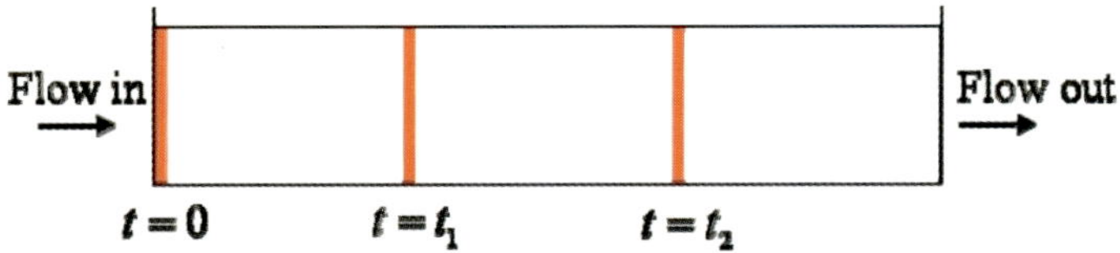

Figure 4.9: Schematic representation of tracer location at different times in a plug flow system.

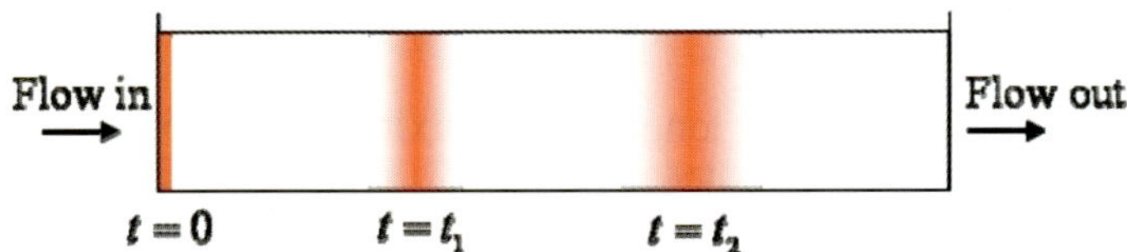

Figure 4.10: Pulse input tracer at different times in a dispersed plug flow system.

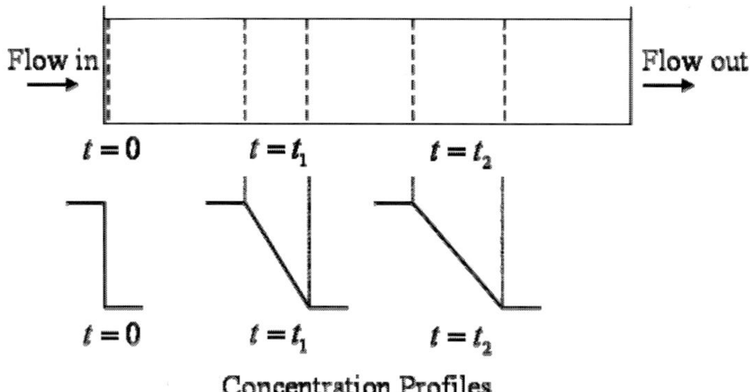

Figure 4.11: Step input tracer at different times for a dispersed plug flow system.

This model characterizes the longitudinal mixing by a one-dimensional equation similar to the Fick's law of diffusion. The constant of proportionality in this equation is known as the eddy diffusivity or dispersion coefficient, *De*. The dispersion of tracer in a continuous flow system results from the eddy diffusion and bulk flow. Thus, the unsteady state concentration may be expressed as

$$\frac{\partial c}{\partial t} = De\frac{\partial^2 c}{\partial x^2} - U\frac{dc}{dx} \tag{4.13}$$

where U is bulk flow velocity of the fluid, and x is the distance coordinate which extends from 0 to the length of the vessel, L. Equation (4.13) can be written as

$$\frac{\partial c}{\partial \theta} = \frac{De}{UL}\frac{\partial^2 c}{\partial y^2} - \frac{\partial c}{\partial y} \tag{4.14}$$

where y is the dimensionless distance and is equal to x/L and $\theta = t/\bar{t} = tU/L$.

In Eq. (4.14), a dimensionless group, De/UL, is called the *vessel dispersion number* or simply the *dispersion number*. This is the inverse of the Peclet number, and is a measure of the extent of the longitudinal dispersion. The eddy diffusivity, De, is not a molecular property of the fluid, but rather depends on the flow conditions of the system. The dispersion number describes interdiffusion from a packet of fluid termed an eddy, over a distance called the eddy length. This movement of eddies in the fluid constitutes mixing. The dispersion number represents a ratio of the amount of fluid transported by eddy diffusion to the amount of fluid transported by the bulk flow. When $De/UL \rightarrow 0$, dispersion is negligible and the plug flow prevails, whereas if $De/UL \rightarrow \infty$, dispersion is at the maximum possible and the well-mixed flow prevails.

For flow characterization of a system, an experimentally obtained residence time distribution curve (F-curve or C-curve) is compared with a family of theoretically obtained curves, and the one that most closely fits the experimental curve is selected. The value of the dispersion number for this curve describes the flow characteristics of the system.

4.4.1.1 *Step input and F-curve*

The solution of Eq. (4.13) for the step input can be obtained by using the following initial and boundary conditions.

The initial conditions (at $t = 0$) are

$$c = 0 \text{ at } x > 0 \tag{4.15}$$

$$c = c_i \text{ at } x = 0 \tag{4.16}$$

The boundary conditions are

$$c = c_i \text{ at } x = 0 \tag{4.17}$$

$$c = c \text{ at } x = L \tag{4.18}$$

The solution of Eq. (4.13) is

$$F \equiv \frac{c}{c_i} = \frac{1}{2}\left[1 - erf\left(\frac{1-\theta}{2\left(\dfrac{De}{UL}\right)^{1/2}}\right)\right]$$

(4.19)

Equation (4.19) is a relationship between dimensionless time, θ and dimensionless concentration, F, for a given value of a dispersion number. Figure 4.12 shows F-curves for different values of dispersion numbers. The experimentally obtained F-curve can be matched with a curve to obtain the value of the dispersion number for the flow system. The F-curves in Fig. 4.12 are not very sensitive to changes in the dispersion number. Thus, generally a C-curve is used to characterize flow systems.

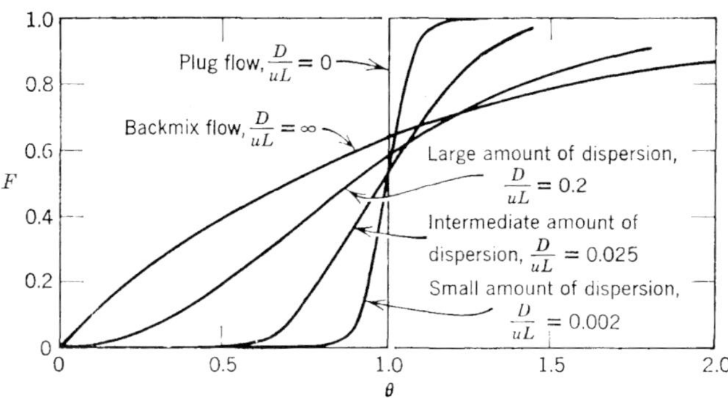

Figure 4.12: F-curves for various dispersion number values as predicted by the longitudinal dispersion model. [Ref. 1]

4.4.1.2 *Pulse input and C-curve*

The solution of Eq. (4.13) depends upon the boundary conditions imposed by the vessel. An analytical solution of Eq. (4.13) has not been found for any other boundary conditions except for one case, the open vessel. However, the mean and variance can be determined in all cases, and these can be used to evaluate the value of the dispersion number for a real system. Many of these values are reported by Van der Laan [2].

4.4.1.3 *Closed vessel*

As shown in Fig. 4.13, a closed vessel has a finite length, L, in which no tracer material moves into or out of the vessel boundaries by dispersion. In other words, the eddy diffusivity at the inlet and outlet is zero. The solution of Eq. (4.13) with the boundary conditions of a closed vessel can be obtained numerically. The mean, $\bar{\theta}$, and variance, σ^2, of this family of curves is found to be

$$\bar{\theta} = 1 \tag{4.20}$$

$$\sigma^2 = 2\left(\frac{De}{UL}\right) - 2\left(\frac{De}{UL}\right)^2\left[1 - \exp\left(-\frac{UL}{De}\right)\right] \tag{4.21}$$

For flow characterization purposes, a continuous casting tundish is a closed vessel, as the tracer at the inlet and the outlet does not move in or out of the tundish by dispersion, and thus, Eqs. (4.20) and (4.21) are applicable. For a closed vessel, the mean residence time, $\bar{\theta}$, of the entire curve is always unity.

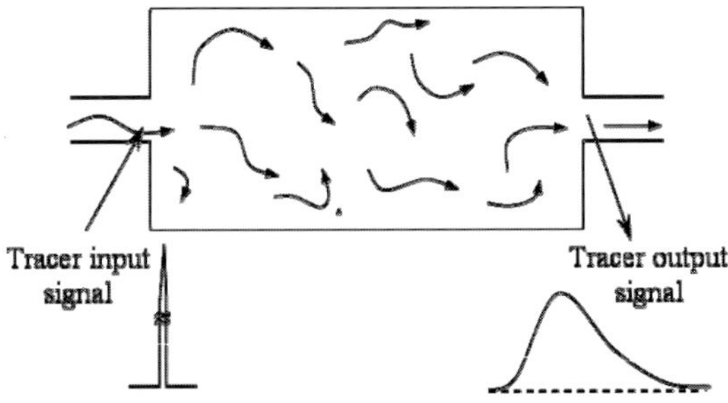

Figure 4.13: Dispersion in a closed system.

4.4.1.4 *Open vessel*

As shown in Fig. 4.14, an open vessel has no discontinuity at the location of tracer injection or the tracer concentration measurement. A selected experimental length in an infinite vessel is termed an open vessel. Thus, the eddy diffusivity at the injection and measurement points is not zero in an open vessel. This is the only situation where the equation for the C-curves has been derived analytically, and is given by

$$C = \frac{1}{2\sqrt{\pi\theta\left(\dfrac{De}{UL}\right)}} \exp\left[-\frac{(1-\theta)^2}{4\theta\left(\dfrac{De}{UL}\right)}\right]$$ (4.22)

The mean and variance for this family of curves are found to be

$$\overline{\theta} = 1 + 2\left(\frac{De}{UL}\right)$$ (4.23)

$$\sigma^2 = 2\left(\frac{De}{UL}\right) + 8\left(\frac{De}{UL}\right)^2$$ (4.24)

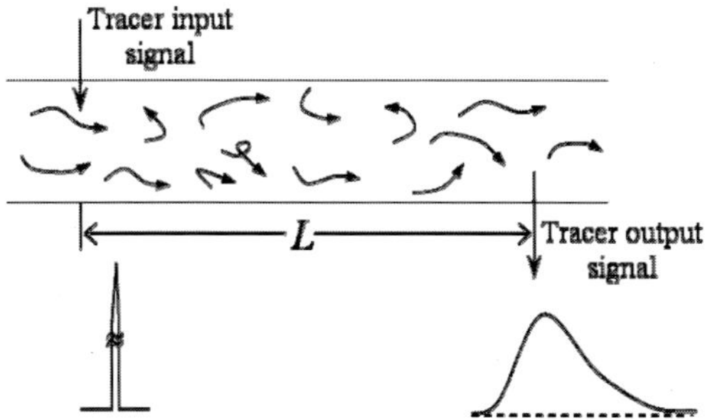

Figure 4.14: Dispersion in an open system.

4.4.1.5 *Dispersion model for a small extent of dispersion (for a closed system)*

For small dispersion number values, the C-curve has a Gaussian type of RTD. In this case, an approximate solution of Eq. (4.13) is not difficult, and is given by

$$C = \frac{1}{2\sqrt{\pi\left(\dfrac{De}{UL}\right)}} \exp\left[-\frac{(1-\theta)^2}{4\left(\dfrac{De}{UL}\right)}\right] \tag{4.25}$$

The mean and variance are given by

$$\bar{\theta} = 1 \quad \text{and} \quad \sigma^2 = 2\frac{De}{UL} \tag{4.26}$$

The error in the value of De/UL obtained using Eq. (4.25) instead of the more exact and complex solutions of Eq. (4.13) increases with the increasing dispersion number. The maximum error in the estimate of De/UL using Eq. (4.25) is less than 0.5% when $De/UL < 0.001$ and is less than 5% when $De/UL < 0.01$ [1].

Figs. 4.15 and 4.16 show the predicted C-curves for large and small values of the dispersion number for a closed system. As can be seen from these figures, C-curves are very sensitive to changes in the dispersion number.

4.4.1.6 *Mean and variance of the residence time distribution*

The mean value of the RTD curve is defined as

$$t_{mean} = \frac{\displaystyle\int_0^\infty tc\,dt}{\displaystyle\int_0^\infty c\,dt} \tag{4.27}$$

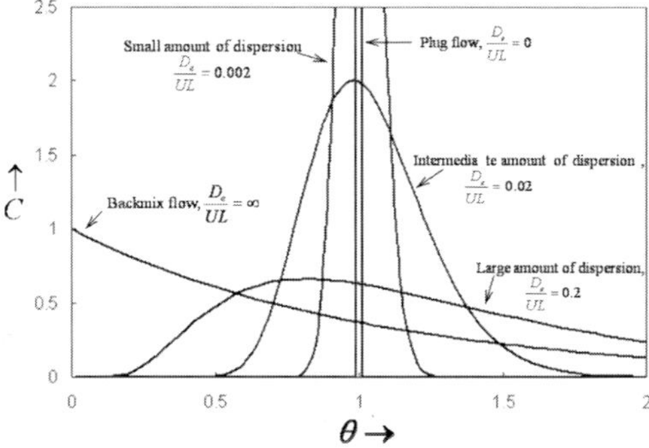

Figure 4.15: C-curves for large values of the dispersion number for a closed system. [Ref. 1]

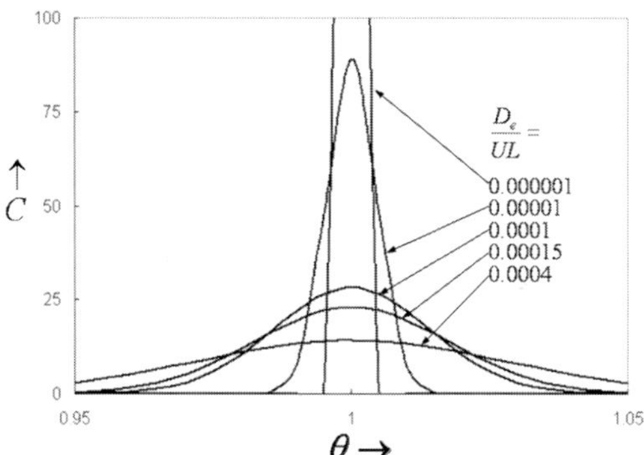

Figure 4.16: C-curves for small values of the dispersion number for a closed system. [Ref. 1]

If the concentration is measured at equal time intervals, Δt

$$t_{mean} = \frac{\sum_i t_i c_i \Delta t}{\sum_i c_i \Delta t} = \frac{\sum_i t_i c_i}{\sum_i c_i} \qquad (4.28)$$

The statistical variance which measures the spread of the residence time distribution about the mean is defined as

$$\sigma_t^2 = \frac{\int_0^\infty (t - t_{mean})^2\, c\,dt}{\int_0^\infty c\,dt} \tag{4.29}$$

Again, for concentration measurements at equal time intervals

$$\sigma_t^2 = \frac{\sum_i (t_i - t_{mean})^2\, c_i \Delta t}{\sum_i c_i \Delta t} = \frac{\sum_i t_i^2 c_i}{\sum_i c_i} - t_{mean}^2 \tag{4.30}$$

From a C-curve, the dimensionless mean of the residence time distribution can be calculated directly by

$$\overline{\theta} = \frac{\int_0^\infty \theta\, C\,d\theta}{\int_0^\infty C\,d\theta} \tag{4.31}$$

and the dimensionless variance is given by

$$\sigma^2 = \frac{\int_0^\infty (\theta - \overline{\theta})^2\, C\,d\theta}{\int_0^\infty C\,d\theta} \tag{4.32}$$

Again, for the measurements at equal time intervals

$$\overline{\theta} = \frac{\sum_i C_i \theta_i}{\sum_i C_i} \tag{4.33}$$

and

$$\sigma^2 = \frac{\sum_i \theta_i^2 C_i}{\sum_i C_i} - \overline{\theta}^2 \tag{4.34}$$

4.4.2 Tanks-in-series model

A series of well mixed tanks can give tracer response curves that are somewhat similar in shape to those found for the dispersion model. Thus, either type of model could be used to correlate experimental tracer data.

The C-curve for one well-mixed tank is given by the equation

$$C = e^{-\theta} \tag{4.9}$$

The C-curve for n well-mixed tanks in a series is given by the equation

$$C = \frac{n^n \theta^{n-1} e^{-n\theta}}{(n-1)!} \tag{4.35}$$

The variance of these curves is given by

$$\sigma^2 = \frac{1}{n} \tag{4.36}$$

Thus, the experimental C-curve data can be matched with the theoretical C-curve obtained using an appropriate value for the number of tanks, n. The number of tanks may be a non-integer. This is also a one-parameter model, with the single parameter being the number of tanks. The C-curves predicted for the tanks-in-series model are shown in Fig. 4.17. For a large number of tanks (n), the C-curve becomes increasingly symmetrical. Since both the tanks-in-series model and the dispersion model give the same general shape of C-curve, these two models can be compared. Eqs. (4.21) and (4.36), for a closed vessel, give

$$\frac{1}{n} = 2\left(\frac{De}{UL}\right) - 2\left(\frac{De}{UL}\right)^2\left[1 - \exp\left(-\frac{UL}{De}\right)\right] \tag{4.37}$$

Levenspiel and Bischoff [3] have given other expressions for comparison of these two models. The tanks-in-series model has been extended by various researchers [Refs. 4-6] to deal with complex flow situations in a reactor. These models basically represent the tanks-in-series model with either backflow or interstage circulation between the consecutive tanks. The imperfect mixing in each tank has also been considered. Shah *et al.* [7] have reviewed all these models.

4.4.3 Combined or mixed models

In addition to the models discussed above, there are several models in which the fluid volume in the vessel is assumed to consist of interconnected flow regions, with various types of flow between and around these regions. Such models are called *combined models* or *mixed models*. Three kinds of flow regions may be assumed to exist in a vessel: the plug flow region, the well-mixed region, and the dead region. The simplest type of mixed model, consisting of the three regions mentioned above, has been extensively used for analyzing the melt flow in tundishes. The calculation of the dead region in this model requires some further discussion, and is presented in one of the succeeding sections.

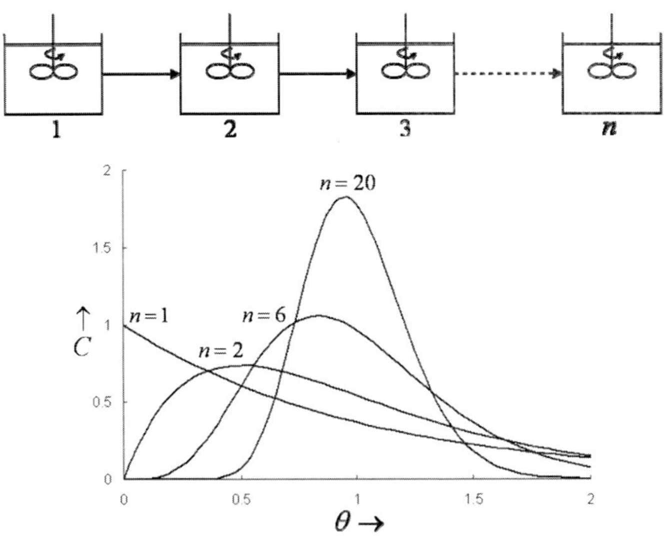

Figure 4.17: C-curves for the tanks-in-series model.

4.4.3.1 *Active region*

Any combination of the plug flow and well-mixed flow regions may be termed an *active region*, as shown in Figs. 4.18 (a) and (b). The order of the two regions is reversed in the two schematic models. The two models give an identical tracer response to a pulse or to any other type of input for a linear system. A linear system is one in which any change in the input or stimulus signal results in a corresponding proportional change in the output or response signal. From the residence time distribution curve shown in Fig. 4.19, the minimum residence time corresponds to the plug volume fraction (V_p / V), and the maximum concentration, C_{max} is equal to the inverse of the well-mixed volume fraction (V / V_m). The equation for the exponential decay curve is given in the Fig. 4.19.

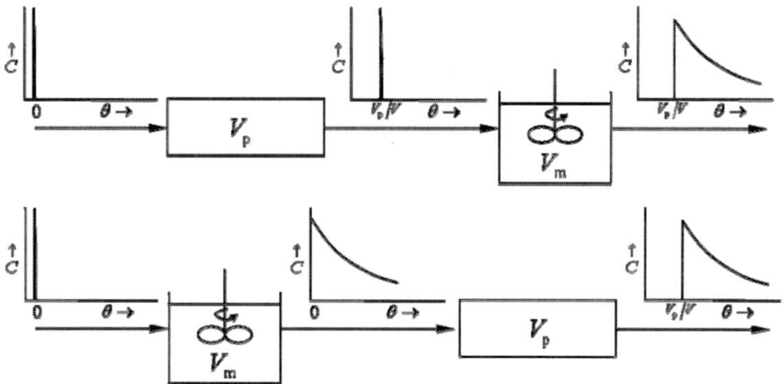

Figure 4.18: (a) and (b) A combined model representing plug volume and mixed volume.

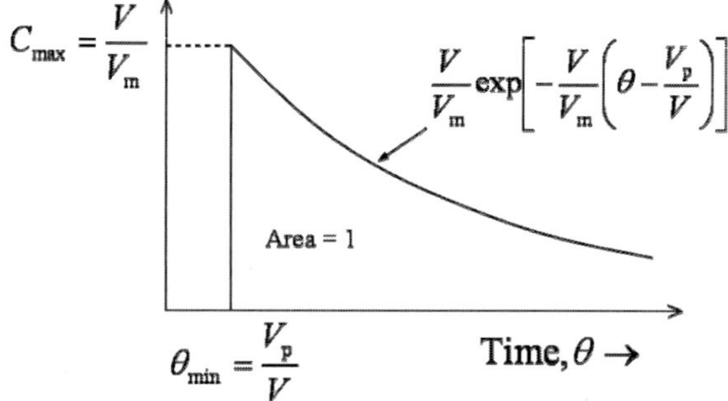

$$\frac{V}{V_m}\exp\left[-\frac{V}{V_m}\left(\theta-\frac{V_p}{V}\right)\right]$$

$$C_{max}=\frac{V}{V_m}$$

Area = 1

$$\theta_{min}=\frac{V_p}{V}$$

Time, $\theta \rightarrow$

Figure 4.19: C-curve for a combined model presented in Fig. 4.18.

4.4.3.2 *Dead region*

Consider a combined model consisting of an active region (plug flow and well-mixed flow) and a dead region. As depicted in Fig. 4.20, the total volume, V, of the system is divided into an active volume of V_a and a dead volume of V_d. The total volumetric flow rate, Q, through the system is also divided into Q_a through the active region and Q_d through the dead region.

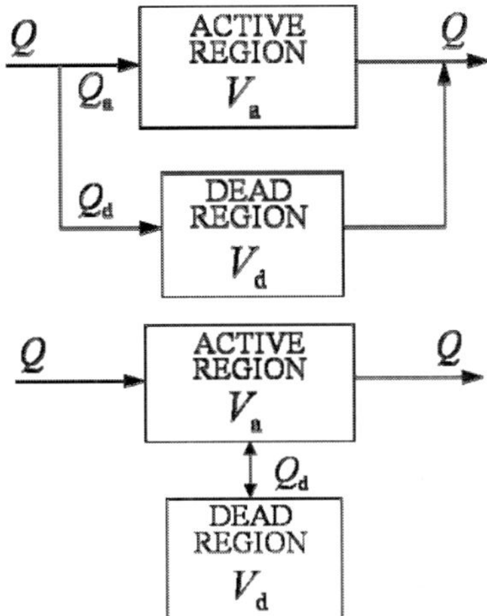

Figure 4.20: Flow through active and dead regions of a combined model.

The dead region may be of one of the following two types. The region may be completely stagnant such that the incoming fluid does not even enter this region. Alternatively, the fluid in this region moves very slowly, and as a result some fluid remains much longer in the vessel. According to the definition by Levenspiel [1], the fluid that stays in the vessel for a period longer than twice the theoretical mean residence time is considered to be the *dead volume*. For a dead region with slowly moving fluid, an experimentally obtained RTD is shown in Fig. 4.21.

Let the mean of the C-curve up to the cutoff point of $\theta = 2$ be $\overline{\theta}_c$, then

$$\overline{\theta}_c = \frac{\text{measured } \overline{t}_c}{\overline{t}} \tag{4.38}$$

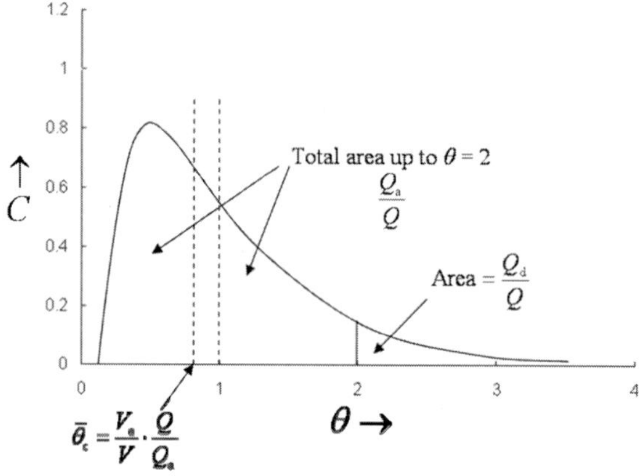

Figure 4.21: Residence time distribution curve for a combined model including a dead region.

$$\bar{\theta}_c = \frac{\bar{t}_c}{\bar{t}} = \frac{V_a/Q_a}{V/Q} = \frac{V_a}{V}\frac{Q}{Q_a} \tag{4.39}$$

$$\frac{V_a}{V} = \frac{Q_a}{Q}\bar{\theta}_c \tag{4.40}$$

Thus, the dead volume fraction becomes

$$\frac{V_d}{V} = 1 - \frac{Q_a}{Q}\bar{\theta}_c \tag{4.41}$$

The term Q_a/Q represents an area under the C-curve from $\theta = 0$ to 2. With the presence of a dead region, the measured average dimensionless residence time is

$$\bar{\theta}_c < 1 \tag{4.42}$$

If the dead region is completely stagnant so that the flowing fluid does not enter or leave the region, then the volume of the system through which the fluid flows in the system is effectively reduced to V_a (or Q_a / Q is unity in Eq. (4.41)). Thus, the dead volume fraction is

$$\frac{V_d}{V} = 1 - \overline{\theta_c} \tag{4.43}$$

The dead volume fraction with the stagnant volume is given by Eq. (4.43), while the dead volume fraction with the slowly moving fluid is given by Eq. (4.41).

In addition to the flows described above, the following kinds of flows may also be present in the combined models: bypass flow, recycle flow, and cross flow. Some models incorporating these flows can be found in Ref. [1].

4.4.3.3 *Application of a combined model to melt flow in tundish*

The combined model has been extensively used to analyze the volumes of plug flow, well-mixed flow, and dead regions in continuous casting tundishes. The typical experimental C-curve obtained in water model studies or in an actual tundish shows an extended tail after the time $\theta = 2$, which indicates the existence of the slow moving flow through the dead regions. As shown schematically in Fig. 4.22, there may exist dead regions with slow moving fluid downstream from the dams and weirs, or near the end wall. The fluid in these regions is constantly interchanged with the main flow (in active volume) of the tundish. These regions should not be considered as stagnant dead volume regions. As stated above, the dead volume fraction in tundishes is given by Eq. (4.41).

The published literature on tundish modeling shows that the combined model has been misinterpreted and used incorrectly. There are two of these incorrect approaches used by researchers. The first one, which has been most widely used [e.g. Refs. 8, 9], is the use of Eq. (4.43). Thus, the model assumes that the area under the curve from the time $\theta = 2$ to ∞ in Fig. 4.21 is zero. In most tundish flow systems, the RTD curve extends much beyond $\theta = 2$. This area represents the fraction of the volumetric flow rate through the dead regions (Q_d/Q). In water modeling experiments of typical tundish designs, there is always an exchange of liquid between the main flow (active volume) and the so-called dead volume regions. A completely stagnant volume assumption that there is no fluid exchange between the active and dead volume, would lead to some error in calculation.

In the second approach [e.g. Ref. 10], which has been used less commonly, the dead volume fraction is considered to be equal to the area under the curve from the time $\theta = 2$ to ∞. This area, in fact, is the fractional volumetric flow rate through the active region (Q_d/Q). No publication in the open literature has been found where Eq. (4.41) has been used for calculation of the dead volume. Table 4.1 gives the estimated error in the dead volume calculation using Eq. (4.43). It is assumed here that the dead volume without any flow is 10% of the total volume. Dead volumes with 1, 5, and 10% fractional flow through the dead regions have been given in Table 4.1. An error as high as 90% arises with 10% flow though the dead region.

Table 4.1: Estimated error in dead volume calculation by ignoring the Q_d/Q.

Q_d/Q	V_d/V	Error in V_d/V
0	0.1	0
0.01	0.109	9%
0.05	0.145	45%
0.1	0.19	90%

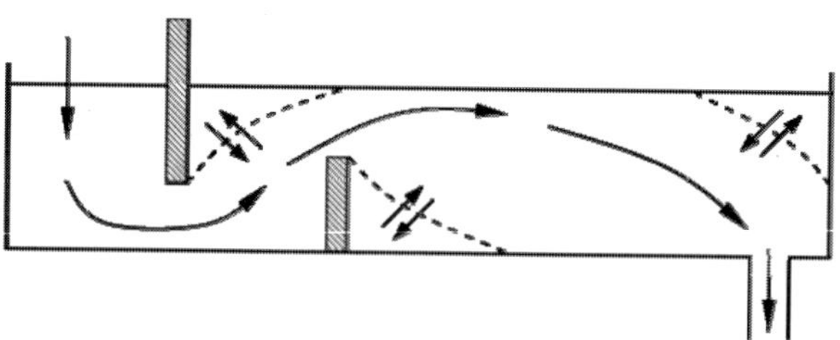

Figure 4.22: Representation of melt flow in a tundish having dead volume exchange of liquid with the active volume.

After calculation of the dead volume, it remains to evaluate the plug flow and well-mixed flow volumes in the tundish. For this, two approaches are suggested. In the one normally used by many researchers, the dimensionless time, θ_{min}, for the first appearance of tracer at the tundish exit is equal to the fractional plug flow volume. The rest of the flow system constitutes the well-mixed volume. Thus, the following equations may be used:

$$\frac{V_p}{V} = \theta_{min} \qquad (4.44)$$

$$\frac{V_m}{V} = 1 - \frac{V_p}{V} - \frac{V_d}{V} \qquad (4.45)$$

By the second approach, the variance of the RTD, σ^2, is calculated. Eq. (4.21) gives a relationship between the variance and the dispersion number (De/UL). Thus, the dispersion number for a given configuration may be calculated. The value of the dispersion number provides the deviation from the ideal plug flow. The dispersion number is zero for plug flow and infinity for a well-mixed flow system. Thus, a higher dispersion number indicates a smaller plug flow region.

4.4.3.4 *An example of tundish melt flow characterization*

Consider a typical Residence Time Distribution experiment for melt flow in a tundish, in which a typical RTD curve extends well beyond the dimensionless time $\theta = 2$. This C-curve, shown in Fig. 4.21, has been analyzed by a combined model for various flow regions in the system.

Area under the entire curve,

$$\sum_{\theta=0}^{\infty} C_i \Delta\theta = 1 \qquad (4.46)$$

Mean residence time for the entire curve,

$$\bar{\theta} = \frac{\sum\limits_{\theta=0}^{\infty} C_i \Delta\theta}{\sum\limits_{\theta=0}^{\infty} C_i} = 1 \tag{4.47}$$

Mean residence time up to $\theta = 2$,

$$\bar{\theta}_c = \frac{\sum\limits_{\theta=0}^{2} C_i \Delta\theta}{\sum\limits_{\theta=0}^{2} C_i} = 0.857 \tag{4.48}$$

Area under the curve up to $\theta = 2$,

$$\frac{Q_a}{Q} = \sum\limits_{\theta=0}^{2} C_i \Delta\theta = 0.9134 \tag{4.49}$$

From Eq. (4.41), the dead volume fraction,

$$\frac{V_d}{V} = 1 - \frac{Q_a}{Q}\bar{\theta}_c = 0.217 \tag{4.50}$$

In this example, the flow rate through the dead region is estimated to be about 9%, and the dead volume calculated from Eq. (4.43) (by neglecting the Q_a/Q volume) is 14.3%. The correct dead volume is 21.7%, which is about 50% more than that predicted by Eq. (4.43). Any change in the tundish configuration, such as the use of different flow control devices, change of total flow rate, or change in depth of liquid in the tundish, could change both the value of $\bar{\theta}_c$ and the Q_a/Q.

4.5 Concluding Remarks

This chapter discusses details of fluid flow characterization as related to the melt flow in tundish. Stimulus-response technique is described, and various models to characterize the residence time distribution of fluid are presented. Application of the combined model for characterizing melt flow in tundish is explained by an example.

References

1. O. Levenspiel, *Chemical Reaction Engineering*, Wiley International, New York, 1972.
2. E. Th. Van der Laan, *Chemical Engineering Science*, 1958, 7, 187.
3. O. Levenspiel and K. B. Bischoff, *Ind. Eng. Chem.*, 1963, 4, 95.
4. H. A. Dean, *Soc. Petrol. Eng. J.*, 1963, 3, 49.
5. C. J. Hoogendoorn and J. Lips, *Can. J. Chem. Eng.*, 1965, 43, 125.
6. J. C. Mecklenburgh, *Trans. Inst. Chem. Engrs.*, 1974, 52, 180.
7. Y. T. Shah, G. J. Stiegel, and M. M. Sharma, *A. I. Ch. E. J.*, 1978, 24, 369.
8. F. Kemeny, D. J. Harris, A. McLean, T. R. Meadowcroft, and J. D. Young, *Proc. Steelmaking Conf.*, ISS, 1981, 2, 232.
9. J. Knoepke and J. Mastervich, *Proc. Steelmaking Conf.*, ISS, 1986, 69, 777.
10. L. K. Chiang: *Proc. Steelmaking Conf.*, ISS, 1992, 75, 437-450.

Chapter 5

Modeling of Melt Flow

5.1 Introduction

Chapters 3 and 4 were devoted to the discussion of fluid flow fundamentals and flow characterization. In this chapter, the application of these concepts and techniques to melt flow studies in tundishes will be discussed. The importance of proper melt flow in continuous casting tundishes for production of clean steel is well recognized. To achieve the desired flow characteristics, a good tundish design and optimum volumetric flow rate of liquid metal are necessary prerequisites. In many cases the use of flow control devices, such as dams, weirs, baffles with holes, pour pads, etc., are found to improve the flow characteristics in the tundish. To assess the effectiveness of these devices and to optimize tundish design, researchers have simulated melt flow and inclusion removal aspects by physical and/or mathematical models before actually using the design in industrial production. The ultimate design objective is to remove as many inclusions as possible and not to create new inclusions during melt flow through the tundish, and thus, to cast cleaner steel.

In physical modeling, a low temperature aqueous analog, generally water, is used to represent molten metal in a tundish. Water flow in a transparent model tundish can be used to observe melt flow physically taking place in an actual tundish. The other method for such studies, as stated above, is mathematical modeling in which melt flow in a tundish is represented by the turbulent Navier Stokes equation. The solution of this equation with an appropriate set of boundary conditions provides detailed information about the velocity and turbulence fields in the system. This information may be further used to analyze various aspects of melt flow in tundishes. With increasing computing power at low cost and the development of more user-friendly commercial software packages, mathematical modeling is gaining popularity and is expected to grow more in the near future. In spite of all the commercial software packages' claims of user-friendliness, however, users must be highly trained in order to take full advantage of these codes. The results obtained by these software packages, like any other computer program, are dependent on the assumptions and the boundary conditions used in solving them. Thus, any inappropriate boundary condition may lead to erroneous and misleading results. Even now, the proper use of these codes to treat the free surface or multi-phase flow which may exist in a tundish is relatively difficult and requires proper training and familiarity with the software packages. It is therefore recommended that the initial mathematical model results must be validated by actual experiments, such as water modeling or industrial trials.

5.2 Physical Modeling

In physical modeling, a full or reduced scale tundish model may be designed based on appropriate similarity criteria (described later) in which the flow of molten metal is simulated by the flow of water. If water flow in the model is a realistic representation of the actual tundish melt flow, it may be used to study various aspects of the melt flow in a tundish, including

(1) liquid splashing from the plunging jet during early stages of tundish filling;
(2) the effect of open stream pouring on air/gas entrainment, slag emulsification and entrainment, and their effects on fluid flow pattern;
(3) free surface wave formation and surface turbulence;
(4) flow visualization in different areas of the tundish with a tracer;
(5) detailed velocity and turbulence measurements in liquid using a probe such as the Laser Doppler velocimeter or Hot Filament anemometer;

(6) study of Residence Time Distribution (RTD) of the fluid;

(7) simulation of inclusion transport and flotation; and

(8) vortex formation and tundish slag entrainment during draining of the melt from the tundish.

The effects of flow control devices on various aspects of fluid flow can also be easily examined using a water model.

5.2.1 *States of similarity*

For a faithful representation of flow in the model tundish, there should be constant ratios between corresponding quantities in the model and in the actual tundish. For melt flow in tundishes, the states of similarity normally include geometric, kinematic, dynamic, and thermal similarities.

5.2.1.1 *Geometric similarity*

The actual tundish and its model must be geometrically similar. All length dimensions of the model tundish should bear a constant ratio with the corresponding length dimensions in the actual tundish. As shown in Fig. 5.1, each length dimension should obey the following relationship:

$$L_m = \lambda \, L_p \tag{5.1}$$

where subscript m and p refer to the model and prototype (the actual tundish), and λ is known as the length scale factor.

5.2.1.2 *Kinematic similarity*

This similarity dictates that each corresponding fluid element in the two systems should follow a geometrically similar path, and the time intervals between the corresponding events should bear a constant ratio. This ratio is known as the time scale factor.

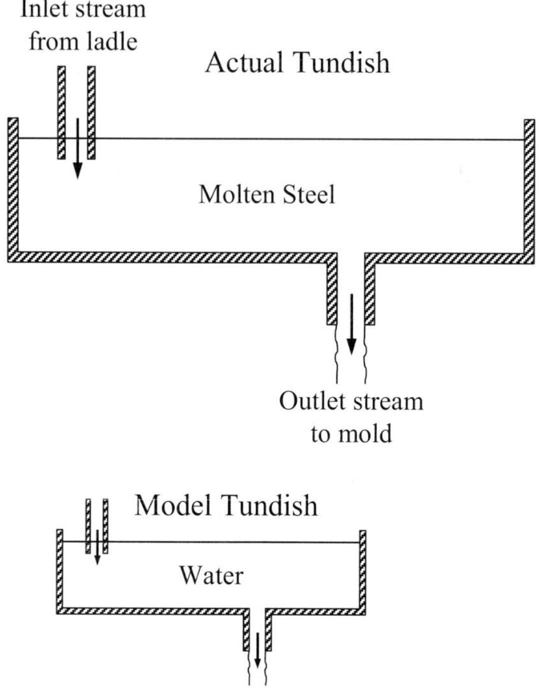

Figure 5.1: Schematic of an actual tundish and a geometrically similar model.

5.2.1.3 *Dynamic similarity*

This similarity deals with the forces acting in the two systems which accelerate or retard the moving element in the system. According to this similarity, forces acting at a corresponding time at a corresponding location in a model tundish should bear a fixed ratio with the forces in an actual tundish. Thus, for example, the important forces which govern liquid flow in tundishes, inertial, viscous, and gravitational, are expressed as:

$$\text{Inertial Force, } F_i = \rho v^2 L^2 \qquad (5.2)$$

$$\text{Viscous Force, } F_\eta = \eta v L \tag{5.3}$$

$$\text{Gravitational Force, } F_g = \rho g L^3 \tag{5.4}$$

Thus at corresponding points at corresponding times in the model and in an actual tundish,

$$\frac{F_{i,m}}{F_{i,p}} = \frac{F_{\eta,m}}{F_{\eta,p}} = \frac{F_{g,m}}{F_{g,p}} = \text{constant}, \lambda_f \tag{5.5}$$

Considering the first two terms in the above relationship will lead to the ratios of inertial to viscous forces in the model and prototype:

$$\frac{F_{i,m}}{F_{i,p}} = \frac{F_{\eta,m}}{F_{\eta,p}} \tag{5.6}$$

$$\left[\frac{\rho v^2 L^2}{\eta v L}\right]_m = \left[\frac{\rho v^2 L^2}{\eta v L}\right]_p \tag{5.7a}$$

$$\text{Re}_m = \text{Re}_p \tag{5.7b}$$

Similarly, ratios of inertial to gravitational forces in the model and actual tundish will give the Froude number equivalence:

$$\frac{F_{i,m}}{F_{i,p}} = \frac{F_{g,m}}{F_{g,p}} \tag{5.8}$$

$$\left[\frac{v^2}{gL}\right]_m = \left[\frac{v^2}{gL}\right]_p \tag{5.9a}$$

$$\text{Fr}_m = \text{Fr}_p \tag{5.9b}$$

5.2.1.4 *Thermal similarity*

In water modeling of tundishes with heat loss, it may become important to consider thermal similarity. This similarity dictates that corresponding temperature differences should bear the same ratio in the model and the actual tundish. Thermal similarity also requires that the rates of heat transfer by conduction, convection, and radiation at a given location at a given time in the model should have a constant ratio with the corresponding rate of heat loss in an actual tundish.

5.2.2 Similarity criteria

The four states of similarity pertinent to tundish modeling are defined above, and if they are maintained between a model and an actual tundish, the model will provide a realistic and correct simulation of the fluid flow phenomenon in the actual tundish. But sometimes it is impossible to maintain all the above stated similarities. In such a case, it is important to understand the impact on the resulting flow in the model tundish which is caused by the ignored aspect(s) of modeling, such as force(s). This issue will now be examined in more detail, as it is probably the most important aspect of physical modeling.

One method of examining the similarity criteria is based on the equation(s) that describe system performance. Consider laminar isothermal flow, which can be described by the Navier Stokes equation given by Eq. (3.8).

$$\rho \frac{Dv}{Dt} = -\nabla p + \eta \nabla^2 v + \rho g \tag{5.10}$$

The body force in the above equation is the gravitational force. Eq. (5.10) may be written in dimensionless form by using the following dimensionless quantities:

$$v^* = \frac{v}{V}; \quad x^* = \frac{x}{L}; \quad t^* = \frac{tV}{L}; \quad p^* = \frac{P - P_0}{\rho V^2}$$

$$\nabla^* = L\nabla; \quad \nabla^{*^2} = L^2 \nabla^2$$

(5.11)-(5.16)

Substitution of the above dimensionless quantities in Eq. (5.10) gives:

$$\frac{Dv^*}{Dt^*} = -\nabla^* p^* + \left(\frac{\eta}{LV\rho}\right)\nabla^{*^2} v + \left(\frac{gL}{V^2}\right)$$

(5.17)

The two dimensionless numbers that appear in parentheses are the inverse of the Reynolds number (Re) and Froude number (Fr). These dimensionless groups also appear in Eqs. (5.7) and (5.9) where it was considered that the important forces governing flow in tundishes were inertial, viscous, and gravitational. Thus, maintaining Reynolds and Froude similarities between a model and an actual tundish would mean having a constant ratio of inertial, viscous, and gravitational forces in flowing fluid in the two tundishes.

The Reynolds similarity between a model (m) and an actual tundish (p) implies that

$$\left(\frac{VL\rho}{\eta}\right)_m = \left(\frac{VL\rho}{\eta}\right)_p$$

(5.18)

Since the kinematic viscosity (η/ρ) of water at room temperature and that of molten steel at 1600 °C is nearly the same (within 10 %), it can be shown that

$$V_m \approx \left(\frac{1}{\lambda}\right)V_p$$

(5.19)

where λ is the length scale factor, and is given by Eq. (5.1). Similarly, the Froude similarity between a model and the actual tundish implies that:

$$\left(\frac{V^2}{gL}\right)_m = \left(\frac{V^2}{gL}\right)_p \qquad (5.20)$$

Thus
$$V_m = \sqrt{\lambda}\, V_p \qquad (5.21)$$

It is obvious from Eqs. (5.19) and (5.21) that satisfying both the Reynolds and Froude similarity criteria in a water model at room temperature is only possible with the use of a full scale model (*i.e.* $\lambda = 1$) by maintaining the same velocity in the model and the prototype. Since in a reduced scale model ($\lambda < 1$) only one of these two criteria can be satisfied. Generally, researchers have chosen the Froude similarity criterion, which uses a reduced velocity in a reduced-scale model in accordance with Eq. (5.21). The effect of ignoring the Reynolds similarity in a reduced scale model will be discussed in the following section.

5.2.3 *Isothermal system*

Consider a water model of melt flow in a tundish with no air or gas entrainment, and in which melt flow is not affected by buoyancy forces caused by any temperature gradients in the melt. Such a flow can be modeled by water at room temperature, referred to as an isothermal system. The importance and validity of Reynolds and Froude modeling criteria can be examined by considering the relevant forces acting in a tundish model system with isothermal flow. The relevant forces involved in such a system are inertial, viscous, and gravitational.

The Reynolds number is the ratio of inertial to viscous forces acting on the fluid. The inertial forces give rise to convective flow or convective momentum transfer, and the viscous forces cause viscous or diffusive momentum transfer. In laminar flow, the molecular viscosity of the fluid causes exchange between the adjacent fluid layers and results in diffusive momentum transfer. However, the flow in tundishes is generally turbulent. In turbulent flows, the diffusive momentum transfer is not only due to the exchange of molecules but also is due to the exchange of eddies over relatively large distances (compared to the molecular mean free path), known as eddy lengths. In such a flow, the contribution of diffusive momentum transfer due to the exchange of eddies may be orders of magnitude greater than the contribution due to the molecular exchange caused by molecular viscosity. Thus, turbulent viscosity is responsible in the same manner for the momentum transfer due to eddy exchange, as the molecular viscosity causes the momentum transfer due to exchange of molecules. The term *effective viscosity* refers to the sum of the molecular and the turbulent viscosities. It is important to point out again (it has earlier been discussed in Chapter 3) that molecular viscosity is a property of the fluid, whereas turbulent viscosity depends on the flow. In turbulent flow, the diffusive momentum transfer depends on the effective viscosity and not on the molecular viscosity. Therefore, the Reynolds number similarity is very important in laminar flows, but becomes less important in turbulent flow modeling.

The Froude number, being the ratio of inertial to gravitational forces, is important where both inertial and gravitational forces are acting on the liquid. Since water in the model tundish is isothermal, the gravitational force does not affect the flow of fluid. Thus, the use of the Froude number similarity criterion is not necessary for modeling of the melt flow alone. However, as is discussed later, the Froude number similarity provides relationships between the actual tundish and its reduced scale model in terms of various scale factors. Also, for modeling of inclusion flotation where lighter inclusion particles rise in the melt, the Froude similarity criterion provides a very convenient method of modeling melt flow and aspects of inclusion coalescence and flotation in a tundish.

Consider the momentum balance equation (Eq. (3.19)) given in section 3.9. For turbulent flow in a tundish with isothermal liquid, the steady state momentum balance equation takes the following form:

$$\frac{\partial\left(\rho v_i v_j\right)}{\partial x_j} = -\frac{\partial P}{\partial x_i} + \frac{\partial}{\partial x_j}\left[\eta_{eff}\left(\frac{\partial v_i}{\partial x_j} + \frac{\partial v_j}{\partial x_i}\right)\right] \tag{5.22}$$

where $\eta_{eff} = \eta + \eta_t$.

In dimensionless form, Eq. (5.22) will yield the following dimensionless number, Re_T, which may be referred to as the *turbulent Reynolds number*.

$$\mathrm{Re}_T = \frac{VL\rho}{\eta_{eff}} \qquad (5.23)$$

Thus, for maintaining flow similarity between two tundishes with turbulent flows, it is important that each term in the dimensionless form of the turbulent momentum balance equation, Eq. (5.22), has the same value in both systems. It is therefore obvious that the flow similarity in two isothermal tundishes can only be maintained if the geometric similarity and the same value of the turbulent Reynolds number are maintained.

To verify these points, Sahai and Emi [1] reported on water and mathematical modeling studies. For water modeling, two laboratory scale tundish models were used. The dimensions of the model tundishes are given in Table 5.1. Model A was 60% in size relative to model B, and the geometric similarity was maintained. In these experiments, water was allowed to flow through the tundish for a sufficient time to ensure the establishment of a steady state. A pulse of dye solution was injected into the incoming stream, and the concentration of the dye at the outlet stream was recorded as a function of time. Plots of the dimensionless concentration against dimensionless time (RTD curve) for different flow rates were obtained.

Results for model tundish A (the smaller model), where water flow rate was varied from 0.13 liter per second (l/s) to 0.76 l/s, are shown in Fig. 5.2. Each curve represents an average of at least two experiments at one flow rate. It can be seen that all RTD curves are very close to each other. Similar experimental results for tundish model B are shown in Fig. 5.3. The volumetric flow rate in this case was varied from 0.13 l/s to 0.88 l/s. The RTD curves for all flow rates, with the exception of the smallest flow rate (0.13 l/s), are even closer than in the previous case.

Table 5.1: Dimensions (in Meters) of the model tundish used in water modeling.

Model Tundish	Length	Liquid Depth	Width	Scale Relative to B
A	0.81	0.24	0.34	0.6
B	1.35	0.40	0.54	1.0

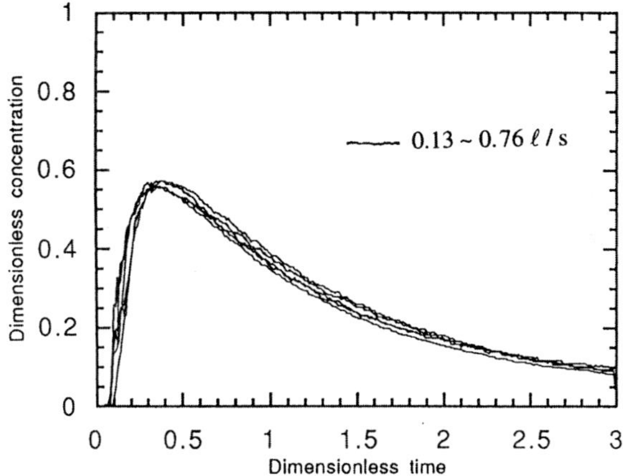

Figure 5.2: Residence time distribution curves in tundish model A (smaller model) for different water flow rates. [Ref.1]

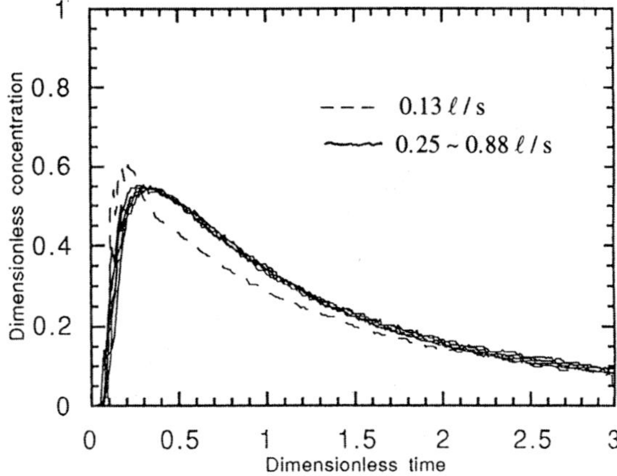

Figure 5.3: Residence time distribution curves in tundish model B (bigger model) for different water flow rates. [Ref. 1]

Averages of all RTD curves for tundish model A and tundish model B (except the smallest flow rate of 0.13 l/s) are plotted in Fig. 5.4. It is obvious that the RTD curves for the two tundish models A and B over a large range of flow rates have very small differences, which may be attributed to experimental error. Thus, the Reynolds and Froude numbers were both varied over a large range, and yet the RTD curve was found to be nearly unchanged. The RTD of the fluid at 0.13 l/s for the bigger tundish is different from all others. This flow rate for the bigger tundish may be so low that the flow regime is no more turbulent. In other words, laminar flow may prevail over a large volume of fluid. In this case, most of the flow volume may be dominated by the laminar viscous forces and may fall into a category where it becomes necessary to satisfy the laminar Reynolds number (based on molecular viscosity).

The liquid flow field and RTD curves for the two laboratory scale model tundishes A and B, and for a full scale tundish were obtained by solving the turbulent three-dimensional Navier Stokes equation. All dimensions of the full scale tundish were three times larger than for tundish model B. Flow in models A and B was solved for water, whereas flows for water and molten steel were solved for the full scale tundish C. Theoretically obtained RTD curves in models A and B for selected flow rates and in a full scale tundish for water and molten steel flow are plotted in Fig. 5.5. It can be seen that all these curves are practically indistinguishable from each other. A comparison of the water modeling and mathematical modeling RTD results also shows a good qualitative agreement, as shown in Figs. 5.4 and 5.5.

Thus, the water and mathematical modeling results have verified that, in isothermal systems where flow is predominantly turbulent, it is not important to match the laminar Reynolds or the Froude number. The analysis also shows that it is important to maintain a constant turbulent Reynolds number, given by Eq. (5.23). For melt flow, the turbulent Reynolds number represents the ratio of the inertial forces resulting in the convective flow and the effective (eddy) viscous forces resulting in the flow of eddies or diffusive flow.

It has been shown by Sahai and Burval [2] that in tundish systems with turbulent flow, the $VL\rho$ term is proportional to the effective viscosity, η_{eff}, over a large range of flow rates and tundish sizes for water and molten steel. In other words, for a given fluid and tundish size, the effective viscosity increases in the same proportion as the increase in fluid velocity, keeping the turbulent Reynolds number constant.

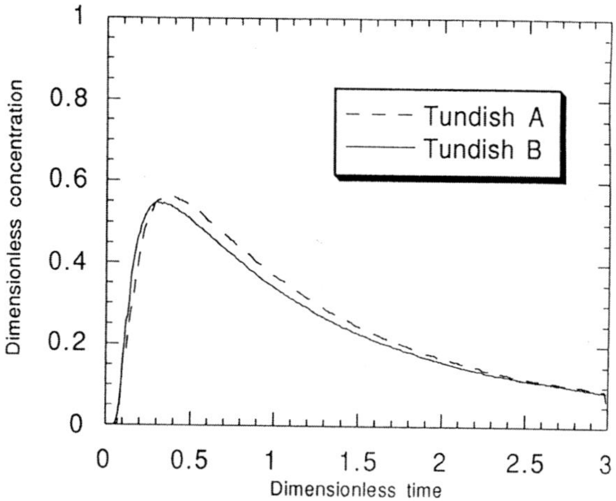

Figure 5.4: Average residence time distribution curves in tundish models A and B over wide range of flow rates. [Ref. 1]

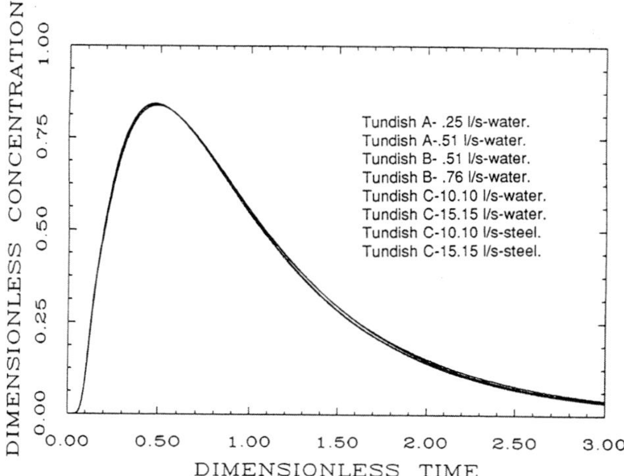

Figure 5.5: Residence time distribution predicted by mathematical models for various flow rates, for different tundish sizes, and for water and molten steel flow. [Ref. 1]

In summary, for isothermal water modeling of melt flow in a tundish which is generally turbulent, the geometric similarity should be maintained and the model may be operated at any flow rate within the turbulent range. The turbulent Reynolds number similarity criterion is naturally satisfied if one operates in the turbulent regime. In a reduced scale model, it is not necessary to satisfy the Froude similarity criterion for melt flow modeling. However, as discussed below, it provides relationships for the parameters relating to the melt flow and inclusion behavior between the model and the prototype tundish. Thus it is common to use the Froude similarity criterion in a reduced-scale water model of a tundish.

5.2.4 *Non-isothermal system*

In actual continuous casting practice, it is possible that melt flow conditions may become non-isothermal. The non-isothermal nature of the flow may be due to heat losses that take place from the top surface and through the walls and bottom of the tundish. More importantly, it is also possible that the temperature of the inlet stream into the tundish from the ladle may vary from heat to heat or with time in the same heat. Thus, there may be situations where the temperature of the inlet stream from the ladle is different from that of the molten steel present in the tundish. These situations are not uncommon in actual casting practice. It has been shown by Chakraborty and Sahai [3] that the fluid flow patterns developed in such a case are quite different from those obtained under isothermal conditions. Thus, it would be useful if water modeling could account for the non-isothermal aspects of the fluid flow phenomenon taking place in a continuous casting tundish. Damle and Sahai [4] examined the necessary modeling criteria for a non-isothermal tundish system. For non-isothermal flows, it is necessary to satisfy the thermal similarity in addition to the geometric and dynamic similarities.

To establish the similarity criteria, the equations governing the flow are converted to dimensionless form. The momentum balance equation for turbulent non-isothermal flow is as follows:

$$\frac{\partial(\rho v_i)}{\partial t} + \frac{\partial(\rho v_i v_j)}{\partial x_j} = \frac{\partial}{\partial x_j}\left[\eta_{eff}\left(\frac{\partial v_i}{\partial x_j} + \frac{\partial v_j}{\partial x_i}\right)\right] - \frac{\partial p}{\partial x_i} + \left(\rho - \rho_{ref}\right)g_i \quad (5.24)$$

The last term in Eq. (5.24) accounts for the buoyancy force per unit volume that arises due to the differences in density throughout the body of the fluid. If β is the coefficient of thermal expansion of the fluid,

$$\beta = -\frac{1}{\rho_{ref}}\left(\frac{\partial \rho}{\partial T}\right)_p \quad (5.25)$$

Therefore, for small variations in density,

$$\rho - \rho_{ref} = -\rho_{ref}\beta\,\Delta T \quad (5.26)$$

In the above equations, ρ_{ref} is a reference density. The Navier Stokes equations may now be rewritten as

$$\frac{\partial(\rho v_i)}{\partial t} + \frac{\partial(\rho v_i v_j)}{\partial x_j} = \frac{\partial}{\partial x_j}\left[\eta_{eff}\left(\frac{\partial v_i}{\partial x_j} + \frac{\partial v_j}{\partial x_i}\right)\right] - \frac{\partial p}{\partial x_i} - \rho_{ref}\beta\,\Delta T g_i \quad (5.27)$$

The temperature distribution within a turbulent flow system is determined by the equation for the conservation of thermal energy.

$$\frac{\partial \rho C_p T}{\partial T} + \frac{\partial(\rho C_p T v_j)}{\partial x_j} = \frac{\partial}{\partial x_j}\left(k_{eff}\frac{\partial T}{\partial x_j}\right) \quad (5.28)$$

In the above equation, k_{eff} is the effective thermal conductivity, which is the sum of the molecular and turbulent thermal conductivities of the fluid. The temperature field is coupled with the velocity field through the buoyancy force term in Eq. (5.27). Eqs. (5.27) and (5.28) together govern the flow dynamics within the system. The boundary condition at the inlet is the ladle stream temperature.

$$T = T_{inlet} \tag{5.29}$$

Eq. (5.27) may be written in dimensionless form as

$$\frac{\partial\left(\rho^* v_i^*\right)}{\partial t^*} + \frac{\partial\left(\rho^* v_i^* v_j^*\right)}{\partial x_j^*} = \frac{\partial}{\partial x_j^*}\left[\frac{\eta_{eff}}{\rho_{ref} VL}\left(\frac{\partial v_i^*}{\partial x_j^*} + \frac{\partial v_j^*}{\partial x_i^*}\right)\right] - \frac{\partial p^*}{\partial x_i^*} - \frac{\beta \Delta TL}{V^2} g_i \tag{5.30}$$

The non-dimensional form of the equation of thermal energy conservation, Eq. (5.28), may be written as

$$\frac{\partial \rho^* T^*}{\partial t^*} + \frac{\partial\left(\rho^* T^* v_j^*\right)}{\partial x_j^*} = \frac{\partial}{\partial x_j}\left[\frac{k_{eff}}{\rho_{ref} C_p VL}\frac{\partial T^*}{\partial x_j}\right] \tag{5.31}$$

where dimensionless temperature, $\;T^* = \dfrac{T - T_0}{T_{inlet} - T_0} = \dfrac{T - T_0}{\Delta T_0}$ $\tag{5.32}$

The dimensionless groups that arise from the above non-dimensional equations are

$$\frac{1}{\mathrm{Re}_T} = \frac{\eta_{eff}}{\rho_{ref} VL} \tag{5.33}$$

$$\mathrm{Tu} = \frac{\beta \Delta T_0 L g_i}{V^2} \tag{5.34}$$

$$\frac{1}{\mathrm{Pr}_T \, \mathrm{Re}_T} = \frac{k_{eff}}{\rho_{ref} C_p VL} \tag{5.35}$$

It has been shown for isothermal system modeling in section 5.2.1 that under turbulent flow conditions, Re_T remains constant. The dimensionless number given by Eq. (5.35) is the inverse of the product of the turbulent Reynolds and turbulent Prandtl numbers. Under turbulent flow conditions, the Prandtl number can be assumed to be constant, and thus the dimensionless group given by Eq. (5.35) remains constant. Since this is the only dimensionless group that arises from the thermal energy equation under the flow conditions being considered, thermal similarity is automatically satisfied.

The dimensionless group given by Eq. (5.34) is therefore the only dimensionless group that remains to be satisfied. This dimensionless number is similar to the modified Froude number. Since this dimensionless group is relevant to the modeling of tundish flows, and all quantities in the definition of the dimensionless group pertain to conditions in the tundish, Damle and Sahai [4] decided to call this number the Tundish Richardson number (Tu). *Tu* denotes the ratio of buoyancy force to the inertial force.

$$\text{Tu} = \frac{\text{Gr}}{\text{Re}^2} = \frac{gL\beta\Delta T_0}{V^2} = \frac{\text{buoyancy force}}{\text{inertial force}} \tag{5.36}$$

The buoyancy forces result from the non-uniform temperature distribution within the tundish. These forces cause the flow profiles within the tundish under non-isothermal conditions to be different from those under isothermal conditions. Keeping Tu constant between the model and the actual tundish should ensure that the ratio of buoyancy force and inertial force in the two systems is the same and thus satisfies the requirements of dynamic similarity between the model and the prototype.

Damle and Sahai [4] tested the validity of the Tundish Richardson number by water and mathematical modeling, in which they changed the temperature difference, ΔT, characteristic length, L, and fluid velocity, V, in a tundish but maintained a constant Tu. They found that the overall fluid flow behavior as reflected by the RTD curves was the same for different conditions. Thus, keeping a constant Tu may be the necessary and sufficient condition for modeling non-isothermal systems.

5.2.5 *Inclusion removal modeling*

Many studies have simulated the removal of non-metallic inclusions in water models by using hollow glass microspheres that were either uncoated or coated by vinyl silane, or polyethylene particles generally in the size range of 20 to 150 microns. These particles were either continuously added to the incoming water or were added as a pulse injection. Several techniques were used for analyzing the number and size of the particles in water leaving the model tundish. These included weighing the particles or observing them under a microscope after filtration, and the use of the Coulter counter for determining the size and number density of the particles. Sahai and Emi [1] discussed the similarity criteria required for appropriate simulation of coalescence and flotation behavior of non-metallic inclusions in tundishes.

5.2.5.1 *Inclusion flotation*

The non-metallic inclusions in steel are lighter than molten steel and thus, rise up to the surface. For the inclusion size range existing in tundishes, it may be assumed that the inclusions rise with the Stokes' velocity. With reference to Fig. 5.6, an inclusion is carried within the fluid at a velocity of $V_{f,p}$, and rises with the Stokes' velocity of $V_{r,p}$. In a model tundish, the corresponding velocities are $V_{f,m}$ and $V_{r,m}$. For the similarity of particle trajectories and hence for proper simulation, it is necessary that the following condition should be satisfied:

$$\frac{V_{f,m}}{V_{f,p}} = \frac{V_{r,m}}{V_{r,p}} \tag{5.37}$$

For this discussion, it is assumed that a full scale water model satisfies the Reynolds and the Froude similarity criteria, and that a reduced scale model is operated with the Froude similarity criterion. In modeling inclusion flotation using a full scale water model, the situation is relatively simple. The lengths, velocities, volumetric flow rates, and corresponding times of different events remain the same in the prototype and in the model. An appropriate material may be used to simulate the inclusions. The size of the model inclusions should be kept the same as the inclusions in steel, and the ratio of the density of the inclusion to the liquid should also be maintained the same in the model tundish and the prototype. This will satisfy the criterion defined by Eq. (5.37), and such modeling experiments are expected to yield information about inclusion flotation which should faithfully represent the real situation.

PROTOTYPE TUNDISH

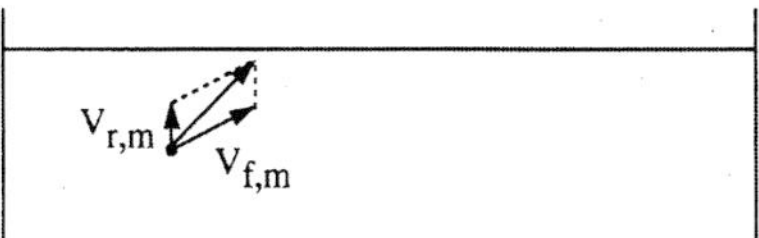

MODEL TUNDISH

Figure 5.6: Schematic representation of inclusion particles and relative velocities in a prototype and model tundishes.

Now imagine that a reduced scale model experiment has been designed based on the Froude similarity criterion. The relationships among velocities and lengths of the model and actual tundish are given by Eqs. (5.21) and (5.1).

$$\text{Velocity, } V_m = \sqrt{\lambda}\, V_p \tag{5.21}$$

$$\text{Length, } L_m = \lambda L_p \tag{5.1}$$

In this experiment, the model inclusions should be selected to have the same density ratio (inclusion to water) as in liquid metal. The important question now is about the size of the inclusions. Based strictly on the Froude similarity, one may propose that the size (diameter) of the model inclusions should be reduced by a factor of λ in accordance with Eq. (5.1). Now examine the question of model inclusion size based on the relative velocities acting on the particle. As stated earlier, for achieving the same rate of flotation/removal in the model as in the prototype, the ratios of the velocities due to bulk flow of liquid and due to Stokes' rise should be same in the two cases. The ratio of velocities due to the liquid bulk flow is given by Eq. (5.21).

$$\frac{V_{f,m}}{V_{f,p}} = \sqrt{\lambda} \tag{5.38}$$

The inclusion's Stokes' rise velocity in the water model and prototype will be

$$V_{r,m} = \frac{2R_{inc,m}^2 g\left(\rho_w - \rho_{inc,m}\right)}{9\eta_w} \tag{5.39}$$

$$V_{r,p} = \frac{2R_{inc,p}^2 g\left(\rho_{st} - \rho_{inc,p}\right)}{9\eta_{st}} \tag{5.40}$$

Thus, for the similarity of the inclusion trajectories in the prototype and its model tundish (given by Eq. (5.37)), the ratio of the two Stokes' rise velocities obtained by dividing Eq. (5.39) by (5.40), should be the same as that due to liquid bulk flow as given by Eq. (5.38). Assuming that the kinematic viscosities of water at room temperature and that of steel at 1600 °C are nearly same,

$$\frac{V_{r,m}}{V_{r,p}} = \frac{R_{inc,m}^2}{R_{inc,p}^2} \frac{\left(1 - \dfrac{\rho_{inc,m}}{\rho_w}\right)}{\left(1 - \dfrac{\rho_{inc,p}}{\rho_{st}}\right)} = \sqrt{\lambda} \tag{5.41}$$

If

$$\frac{\rho_{inc,m}}{\rho_w} = \frac{\rho_{inc,p}}{\rho_{st}} \tag{5.42}$$

$$\frac{R_{inc,m}^2}{R_{inc,p}^2} = \frac{D_{inc,m}^2}{D_{inc,p}^2} = \sqrt{\lambda} \tag{5.43}$$

or

$$R_{inc,m} = \lambda^{0.25} R_{inc,p} \tag{5.44}$$

Thus, the relationship given by Eq. (5.44) should provide similarity among inclusion trajectories in the water model and prototype, and in turn, should correctly simulate the inclusion flotation. For example, in a one-third scale model the inclusion size should be 0.76 times the steel inclusion size. Fig. 5.7 shows the relationship between the steel and water model inclusion sizes for different scale factors.

Sometimes it becomes difficult to find a model inclusion with the same density ratio, *i.e.*

$$\frac{\rho_{inc,m}}{\rho_w} \neq \frac{\rho_{inc,p}}{\rho_{st}} \tag{5.45}$$

Then, from Eq. (5.41)

$$\frac{R_{inc,m}}{R_{inc,p}} = \lambda^{0.25} \left[\frac{\left(1 - \dfrac{\rho_{inc,p}}{\rho_{st}}\right)}{\left(1 - \dfrac{\rho_{inc,m}}{\rho_w}\right)} \right]^{0.5} \tag{5.46}$$

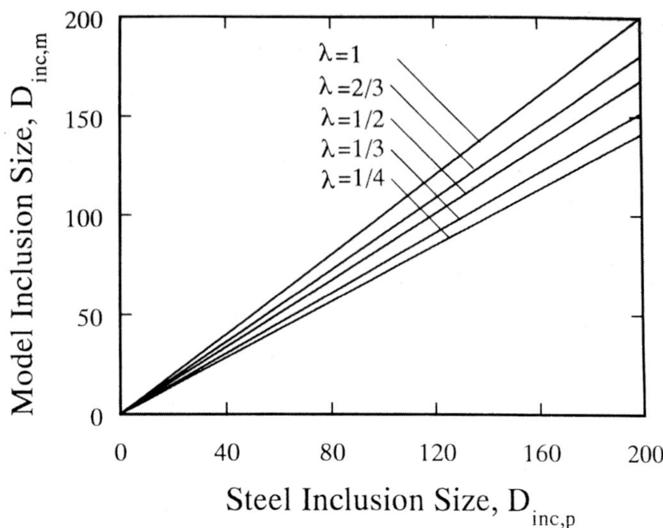

Figure 5.7: Relationship between steel and water model inclusion sizes for different scale factors. [Ref. 1]

One can use Eq. (5.46) for the model inclusion size with different density ratios. Fig. 5.8 shows the relationship between the steel and water model inclusion sizes in a one-third scale model for different values of γ, which is defined by the following equation:

$$\gamma = \left[\frac{\left(1 - \dfrac{\rho_{inc,p}}{\rho_{st}} \right)}{\left(1 - \dfrac{\rho_{inc,m}}{\rho_{w}} \right)} \right] \tag{5.47}$$

Fig. 5.9 shows the ratio of inclusion sizes as a function of different length scale factors for different values of γ. For a one-third scale model ($\lambda = 0.33$), let the density ratio in prototype be 0.6 and that in the model inclusion be 0.8, *i.e.* a γ of 2.

$$\frac{\rho_{inc,p}}{\rho_{st}} = 0.6 \tag{5.48}$$

and
$$\frac{\rho_{inc,m}}{\rho_w} = 0.8 \tag{5.49}$$

Then, from Eq. (5.46) or from Fig. 5.9

$$\frac{R_{inc,m}}{R_{inc,p}} = \frac{D_{inc,m}}{D_{inc,p}} = 1.07 \tag{5.50}$$

In other words, this result simply suggests that if the model's particle to liquid (water) density ratio is greater than that of the corresponding prototype ratio, then the model particle will rise slowly. Hence, a bigger particle is required to compensate for the density difference. This way, one can obtain the inclusion size relationship with a different particle to liquid density ratio. Such particles in a water model would faithfully simulate the flow of inclusions in molten metal in an actual tundish.

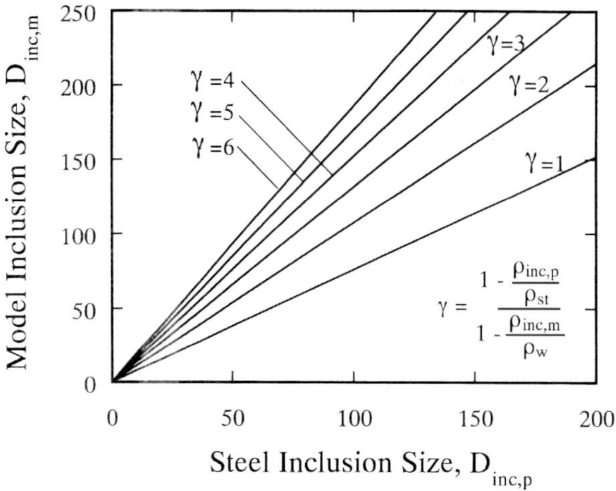

Figure 5.8: Relationship between steel and water model inclusion sizes in a one-third scale model for different values of γ. [Ref. 1]

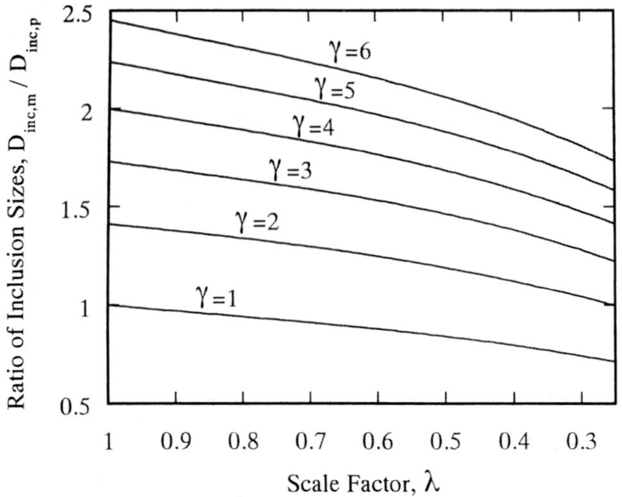

Figure 5.9: Ratio of inclusion sizes as a function of different length scale factors for varying values of γ. [Ref. 1]

5.2.6 *Water modeling procedure*

A full or reduced scale transparent acrylic model of the actual tundish should be designed and built. As discussed in the earlier section, it is important to maintain the geometric similarity between the actual and the model tundishes. The water inlet to the tundish can either be from a model ladle that is kept on top of the model tundish, or water can directly be discharged from a pipe at the desired volumetric flow rate (see Fig. 5.10). Water from the tundish outlet(s) may be either recirculated back to the tundish inlet through a holding tank or directly discharged to a drain. To achieve dynamic similarity, as discussed in sections 5.2.1 and 5.2.2, Tundish Richardson number similarity should be maintained for non-isothermal systems, and the Froude number similarity provides convenient relationships for isothermal systems as follows,

$$\text{Velocity} \qquad\qquad V_m = \sqrt{\lambda}\, V_p \qquad\qquad (5.21)$$

Length $$L_m = \lambda L_p \tag{5.1}$$

Time $$t_m = \sqrt{\lambda}\, t_p \tag{5.51}$$

Volume $$(\text{Volume})_m = \lambda^3 (\text{Volume})_p \tag{5.52}$$

Volumetric flow rate $$Q_m = \lambda^{2.5} Q_p \tag{5.53}$$

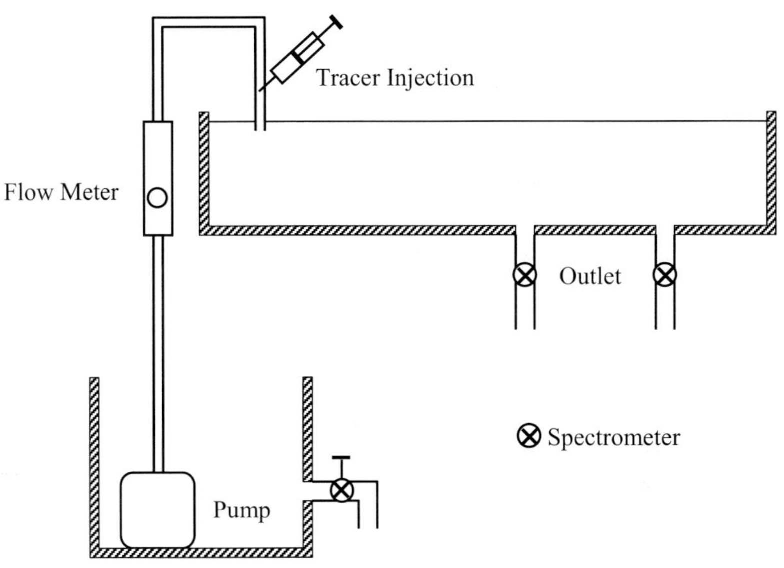

Figure 5.10: Schematic representation of a typical apparatus for tundish water modeling.

When the model setup is ready for experimentation and water at a desired flow rate can be circulated through the system, fluid flow characteristics during steady state and unsteady state operations of the actual tundish can be easily simulated. Various aspects listed in section 5.2 have been studied in water models. The Residence Time Distribution of fluid in tundishes is most commonly studied, and it provides a good indication of the overall flow behavior in a tundish. Some details of the RTD study procedure are discussed here. For RTD studies, a neutrally buoyant solution of tracer is injected into the incoming stream near the inlet. For steady state liquid flow, water should flow through the model for sufficient time to ensure that the steady state flow profile is established before injecting the tracer. For RTD studies, several tracers have been used by researchers, including acid, salt solution, dye solution (such as methylene blue), potassium permanganate, and food coloring dyes. Concentration of the tracer at the outlet is measured using a concentration probe. An on-line pH measuring device, conductivity meter, and colorimeter may be used for acid, salt, and dye concentration measurements, respectively. It is important that the tracer solution have the same density as water, otherwise distorted and misleading information about liquid flow and RTD may result. The effect of tracer density on the flow field is presented in the next section.

If the tracer input is a pulse injection, the resulting time-concentration curve plotted on a dimensionless scale is known as the C-curve. Details of the stimulus-response technique have been discussed in Chapter 4. Dimensionless time, θ, is given by Eq. (4.2), and dimensionless concentration is given by Eq. (4.4). In Eq. (4.4), q/V is the average concentration of tracer dissolved in the tundish volume, V. Sometimes uncertainty in knowing the precise quantity of tracer, q, is a major source of error in RTD studies. Thus, an alternate and more accurate method is to calculate the quantity of the tracer from the area under the C-curve. Plotted on a dimensionless scale, the area under the entire C-curve is always unity. So if the experiment is run long enough and all tracer has exited the tundish, the area under the curve should be normalized to unity. Because there is no ambiguity in the dimensionless time, in normalizing the C-curve to unity, the dimensionless concentration has to be adjusted by changing the value of the average concentration, q/V. Thus, the precise amount of injected tracer can be back-calculated.

The C-curve may be analyzed by any of the models presented in Chapter 4, such as the longitudinal dispersion model, the tanks-in-series model, or a combined model. Thus, the effects of any design changes or use of flow control devices on the mixing characteristics, dead volume, or other aspects can be evaluated.

5.2.7 *Effect of tracer density on melt flow characterization*

Damle and Sahai [5] studied the effect of tracer density on melt flow characterization using mathematical and water modeling techniques. In water modeling they used a neutrally buoyant and saturated KCl solution as tracers in a two-strand tundish. A schematic of the water modeling experimental set-up is shown in Fig. 5.10. Predicted and experimental RTD curves obtained at the two tundish outlets appear in Figs. 5.11 and 5.12, respectively. Curve A represents the RTD at the outlet closer to the incoming stream, and B is at the far outlet. The volume of the tundish was 0.187 m^3, and the volumetric flow rate was 1.9 x 10^{-4} m^3. s^{-1}. They found that the addition of 50 cm^3 of saturated KCl solution (density of 1163 kg. m^{-3}) changed the overall flow profiles with time. The corresponding predicted and experimental RTD curves are shown in Figs. 5.13 and 5.14, respectively. It can be seen that the RTD curves with the KCl solution as tracer are very different from those with the neutrally buoyant tracer. The RTD curves obtained from the mathematical modeling also show the same effect of tracer density.

Copper is quite often used to study RTD in actual tundishes. Damle and Sahai [5] also predicted the RTD curves mathematically for the two-strand tundish with 35 tons of molten steel. Densities of copper and molten steel at 1600 °C were taken as 7599 and 6984 kg. m^{-3}, respectively. Predicted residence time distribution curves with a neutrally buoyant tracer are shown in Fig. 5.15. The predicted RTD curves with the addition of different amounts of copper in this tundish are shown in Fig. 5.16. Thus, it is obvious that the use of tracer of different density than the tundish liquid may significantly distort the RTD curves and may result in a misleading interpretation of the flow. Of course, the distortion depends upon the density difference and the amount of tracer added.

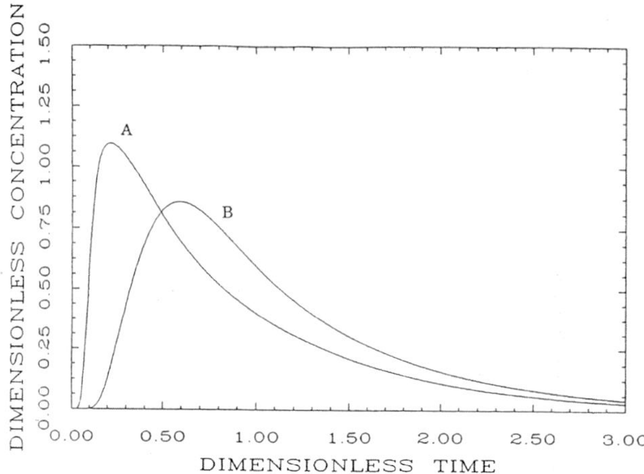

Figure 5.11: Predicted residence time distribution curves obtained for a pulse injection of a neutrally buoyant tracer. [Ref.5]

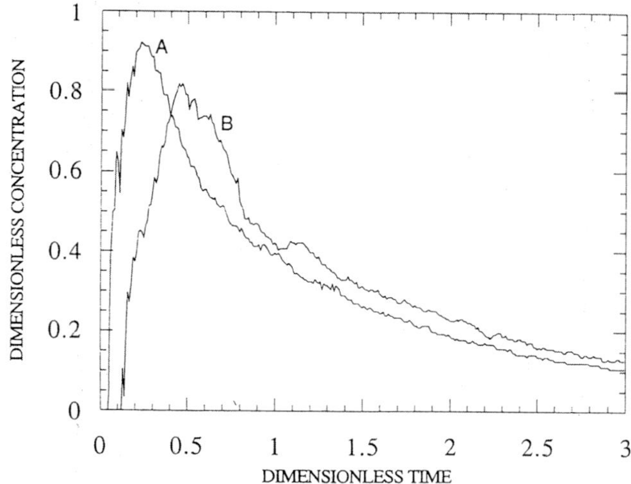

Figure 5.12: Experimental residence time distribution curves obtained for a pulse injection of a neutrally buoyant tracer. [Ref. 5]

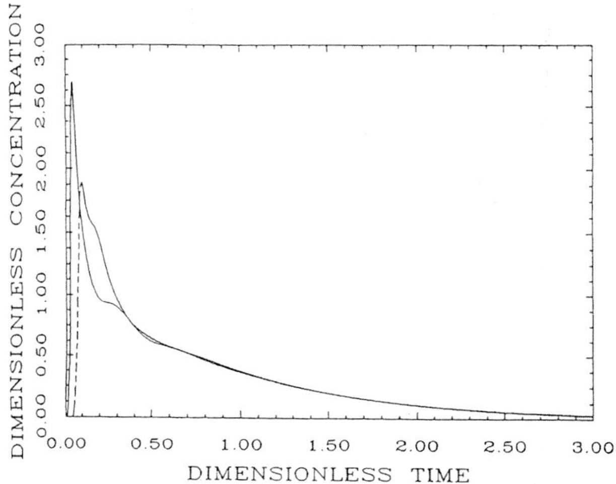

Figure 5.13: Predicted residence time distribution curves obtained for a pulse injection of a saturated KCl solution tracer. [Ref. 5]

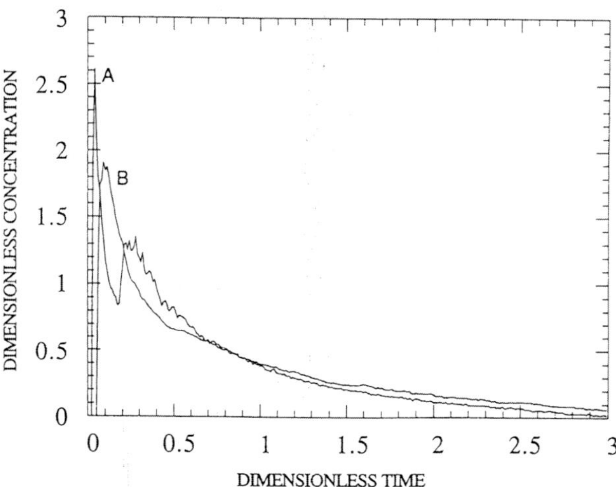

Figure 5.14: Experimental residence time distribution curves obtained for a pulse injection of a saturated KCl solution tracer. [Ref. 5]

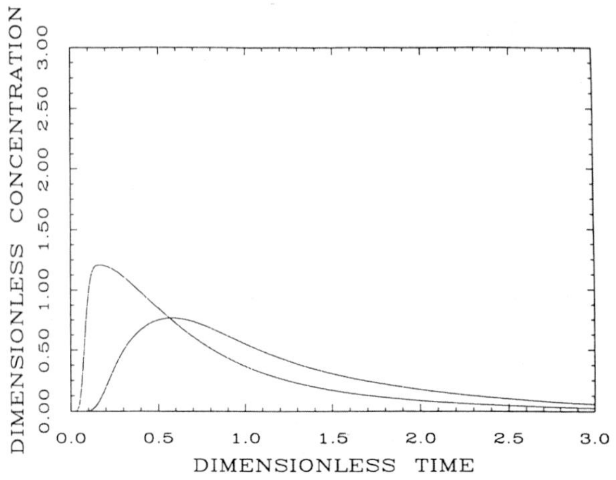

Figure 5.15: Predicted residence time distribution curves obtained for a pulse injection of a neutrally buoyant tracer in a steel tundish. [Ref. 5]

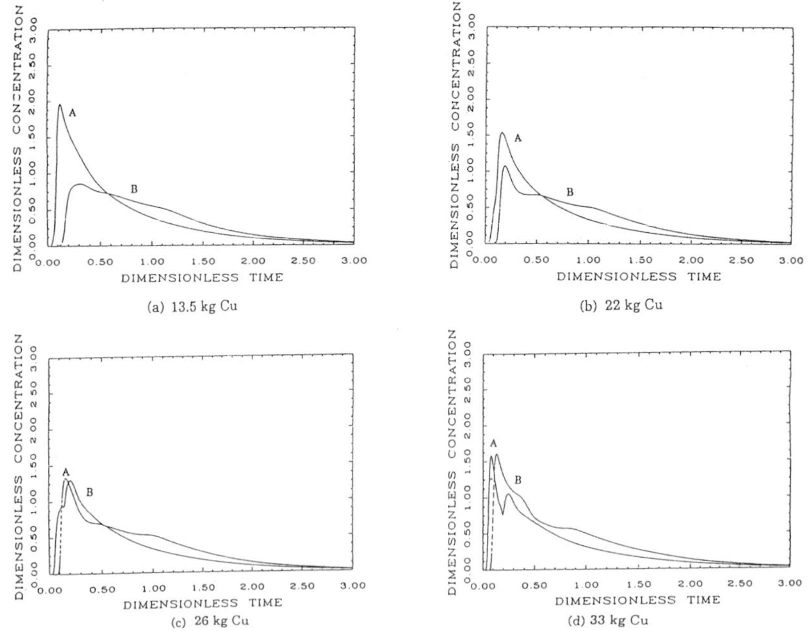

Figure 5.16: Predicted residence time distribution curves obtained for different quantities of copper used as tracer in a steel tundish. [Ref. 5]

5.3 Mathematical Modeling

Water modeling, described in section 5.2, is a very convenient and useful method for gaining a good understanding of the overall flow behavior in a tundish. It can quickly indicate any abnormalities or undesired flow characteristics, such as short circuiting or dead regions in a tundish. Effects of various flow control devices in improving the overall flow characteristics can also be easily studied. In fact, many aspects of flow that can be simulated in a water model are still very difficult and cumbersome to simulate mathematically. On the other hand, mathematical modeling may provide a much more detailed picture of velocity, turbulence, and temperature fields as a function of location and time. Mathematical modeling is also a very useful link between a water model and the actual tundish with molten steel as a liquid. Once the mathematical model results are validated with a reduced- or full-scale water model results, the mathematical model may easily be used for a larger scale tundish with molten metal. Certain configurational changes, such as location and size of the flow control devices, can also be easily incorporated in a mathematical model. Results can also be obtained in a much shorter time compared to the corresponding water modeling experimental results. Thus, the two modeling techniques should be considered as complementary to each other, each with its own advantages and shortcomings.

5.3.1 *Turbulent flow field and heat transfer*

Fluid flow and heat transfer equations are given in Chapter 3. The expanded form of these equations can be found in any standard textbook on transport phenomena. Depending upon the problem, either time-dependent or steady state equations can be solved. For a three-dimensional, turbulent, steady state flow problem in which heat transfer aspects are considered, the following equations are solved:

Equation of continuity (Eq. (3.18))
Momentum balance equations (three equations) (Eq. (3.19))

K and ε equations for evaluating turbulent viscosities (Eqs. (3.26) and (3.27))

Heat transfer equation (Eq. (3.22))

5.3.2 *Boundary conditions and assumptions*

Boundary conditions depend upon the specific problem, but the most commonly used boundary conditions are presented here. Close to solid walls, including any dam or weir, the variation in flow properties is much steeper than within the bulk fluid. Consequently, the momentum (v_1, v_2, and v_3) and scalar (K and ε) transport properties are modeled using wall functions. This method is given in detail by Launder and Spalding [6]. Also, generally no-slip boundary conditions are imposed at the solid walls. At the free surface, which is generally assumed to be flat, and at the symmetry planes, the normal velocity components and the normal gradients of all other variables (momentum and scalar transport properties) are assumed to be zero. At the jet entry, the velocity perpendicular to the free surface may be calculated from the volumetric flow rate and the area of the nozzle. Since, generally, the jet flow is in the highly turbulent range, a flat velocity profile may be assumed. Similar boundary conditions may also be imposed at the outlet nozzle(s).

The incoming liquid jet may be considered to be at a constant temperature. For modeling heat fluxes through the free surface and each wall, heat fluxes are assigned or estimated. Chakraborty and Sahai [7] have described the heat transfer boundary conditions as applicable to tundish modeling in detail.

5.3.3 *Numerical solution procedure*

There are many methods available for discretization of the differential transport equations. The procedure which is outlined here is explained in detail by Patankar [8]. Finite difference equations are derived from the transport equations listed above, using an implicit finite difference procedure referred to as SIMPLE (Semi-Implicit Method for Pressure Linked Equations). There are several commercial computer codes available which can be used to solve these transport equations. Names of some of these computer commercial codes are listed in the next section.

The computational domain is divided into a non-uniform three-dimensional grid. It is generally a good practice to test the dependence of the grid size on the solution of the problem. Of course, a finer grid takes more computer time but generally results in a more accurate solution. Thus, a balance must be struck between the accuracy of solution and computer time.

5.3.4 *Tracer dispersion*

The dispersion of tracer introduced into the liquid flowing through a tundish may be described by the following mass transport equation:

$$\frac{\partial}{\partial t}(\rho C) + \frac{\partial}{\partial x_i}(\rho v_i C) = \frac{\partial}{\partial x_j}\left(\rho D_{eff}\frac{\partial C}{\partial x_i}\right) \tag{5.54}$$

This is a time dependent equation in which C represents the concentration of tracer, with D_{eff} being the effective diffusivity. The effective diffusivity depends upon the fluid flow field. As explained in Chapter 3, D_{eff} is related to effective viscosity through the turbulent Schmidt number as

$$\frac{\eta_{eff}}{\rho D_{eff}} \approx 1 \tag{5.55}$$

The boundary conditions required to solve Eq. (5.54) have to express the physical constraints that all the bounding surfaces are impervious to the tracer. Mathematically, this corresponds to zero flux of tracer at all bounding surfaces. The solution of Eq. (5.54) produces the Residence

Time Distribution curve for the given flow conditions. This may be compared with an experimentally obtained RTD curve.

5.3.5 *Inclusion transport*

The modeling of inclusion transport and removal is quite complex as partly explained in Chapter 2. The treatment presented here is rather simple. The transport equation of particles is similar to Eq. (5.54), where particles are transported convectively with the flow, and diffusively with the effective diffusivity. Since the particles are lighter, they rise with the Stokes' velocity given by Eq. (3.10). This rise velocity is added to the steady state vertical velocity component of liquid flow. Thus particles flow with liquid in the tundish and at the same time rise with Stokes' velocity. The collision and coalescence of the inclusion particles are discussed in Chapter 2. The calculated rate of the coalescence is added to the transport equation as a "sink term." Details of this procedure may be found in Sinha and Sahai [9], who modeled inclusion transport and removal in continuous casting tundishes. In this work, they considered collision and coalescence of inclusion particles, flotation of particles to the free surface where the particles were absorbed by the tundish slag, and particles also stuck to the solid surfaces.

5.3.6 *Commercial software codes*

There are many commercially available software codes available for such calculations. Names of some of the CFD codes are listed below.

(1) FLUENT
(2) FLOW3D
(3) CFX
(4) STARCD

5.4 Case Studies

Selected case studies are presented in this section in which the modeling techniques described in this chapter have been used, and were found to be very beneficial in improvement of the cast product. The quality of the steel cast after implementation of the improved design was better in all these cases.

5.4.1 *Case study # 1*

Harris and Young [10] performed an extensive water modeling study to optimize the performance of Stelco's Lake Erie Works' Caster. They used a full scale water model, and maintained the same volumetric flow rate of water as melt flow in the industrial caster. Thus, the Reynolds and Froude similarities were maintained between the model and the actual tundishes. In their water modeling experiments, they injected potassium permanganate solution as a tracer into the ladle stream. The time for the first appearance of tracer in the mold was recorded, and it was referred to as the minimum retention time. Experiments in the tundish model were conducted with no flow control, then with a weir, and finally with a weir and a dam. They optimized the location and dimensions of the devices by conducting the experimental trials. The optimization was done by maximizing the minimum retention time. The schematic representation of the flow patterns without and with the flow control devices is given in Fig. 5.17. The optimized flow control device(s) designs were used in an actual caster, and a significant reduction in slabs downgraded for aluminate appearance resulted. Reject percentage was 2.9% without any flow control. The reject percentage was reduced to 2.1 % with the weir, and only 0.3% with a permanent weir and dam combination. Thus, the authors concluded that water modeling is a viable tool for tundish design optimization.

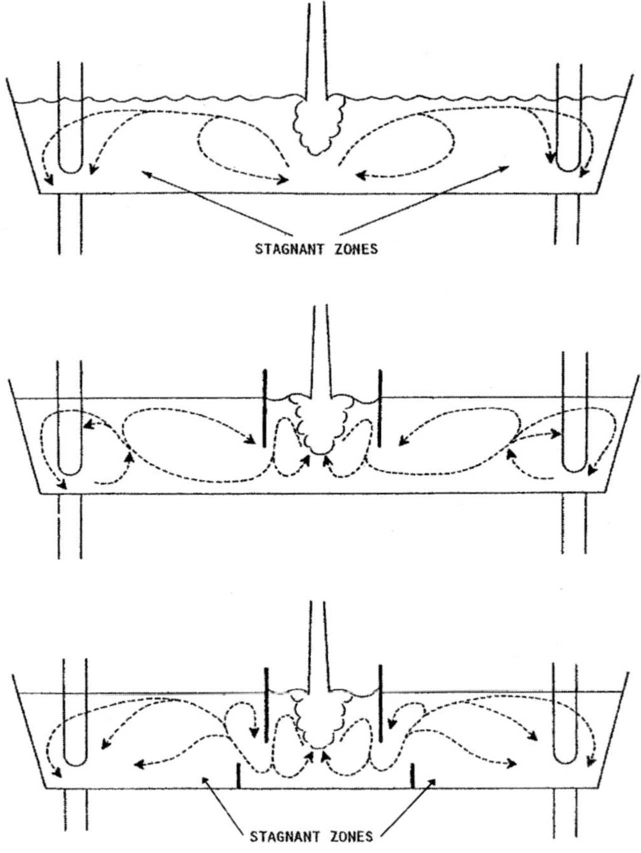

Figure 5.17: Schematic representation of the flow patterns without and with the flow control devices. [Ref. 10]

5.4.2 *Case study # 2*

Lowry and Sahai [11] studied melt flow in a T-shaped 6-strand caster of the Armco Kansas City Works. The effect of a dam, weir, and a baffle with holes on the melt flow was studied. Flow in the tundish was predicted with and without flow control devices using a 3-dimensional computer code for solving the turbulent Navier Stokes equation.

Figs. 5.18 and 5.19 show velocity profiles in symmetrical half of the tundish in various horizontal longitudinal planes. These are horizontal cross sections of the tundish (a) being near the free surface and is widest, and (e) being closer to the tundish bottom so it has the least width. Locations of the three outlets to the mold are visible in the section (e). Sections (a), (b), and (c) show a weir from the top, whereas sections (c) and (d) show a dam sitting the bottom in the short arm of the tundish.

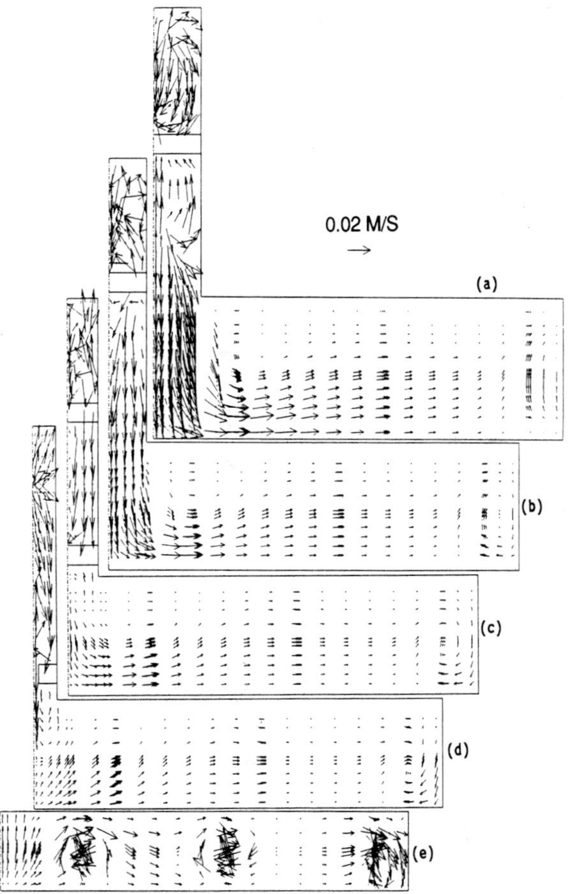

Figure 5.18: Predicted flow fields with a dam and a weir in horizontal transverse planes, (a) being near the free surface and (e) being near the bottom. [Ref. 11]

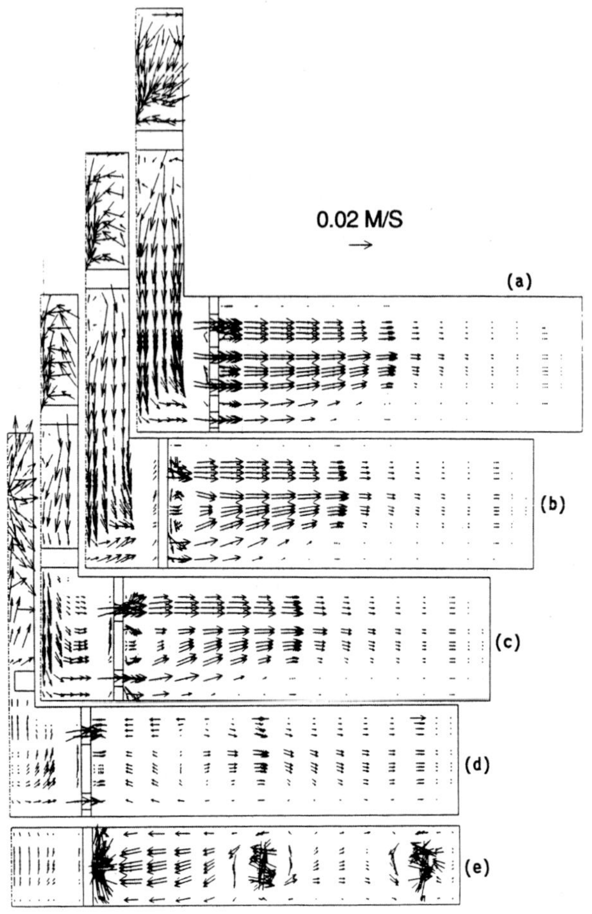

Figure 5.19: Predicted flow fields with a dam, weir, and a baffle in horizontal transverse planes, (a) being near the free surface and (e) being near the bottom. [Ref. 11]

Fig. 5.19 also has a baffle with holes in the long arm of the tundish in addition to the dam and weir in the short arm. Metal from the ladle is poured on the upstream of the weir. The effect of flow control devices on the fluid flow can be seen from Fig. 5.18, which shows that the flow is predominantly towards the outer wall at least in the top two-thirds of the tundish. With the placement of a baffle with holes, the flow has changed considerably. As can be seen in Fig. 5.19, the flow has become more uniform in the entire cross section. Mathematically predicted RTD curves with and without a baffle at the inside and outside strand are shown in Fig. 5.20. The residence time for the inside strand has increased by addition of the baffle.

The RTD study was also made by injecting saturated NaCl solution as a tracer in the water model, and Cu was used as the tracer in the actual tundish. Figs. 5.21 and 5.22 show comparisons of RTDs obtained by water model study, actual tundish measurements with Cu addition, and mathematical model study for the inside and outside nozzles. It has been discussed in an earlier section that Cu and saturated salt solution are not neutrally buoyant tracers, and one should consider the effect of the tracer density difference in using them. Their use as tracer may have distorted the results. It can be seen that the mathematically modeled RTD curve is different than the other two curves in Fig. 5.22. The optimized configuration was implemented in plant practice, and improvements in product quality were observed. In a later study, Lowry and Sahai [12] used hot and cold water in tundish model studies, and showed that the buoyancy effects play an important role in altering the melt flow profiles. Thus, the effects of tracer density and thermal gradients should be carefully considered in characterizing flow in tundishes.

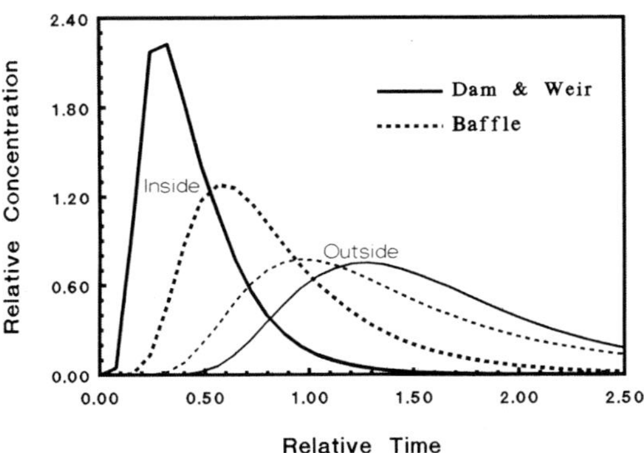

Figure 5.20: Residence time distribution calculated by the mathematical model with and without baffle in the tundish. [Ref. 11]

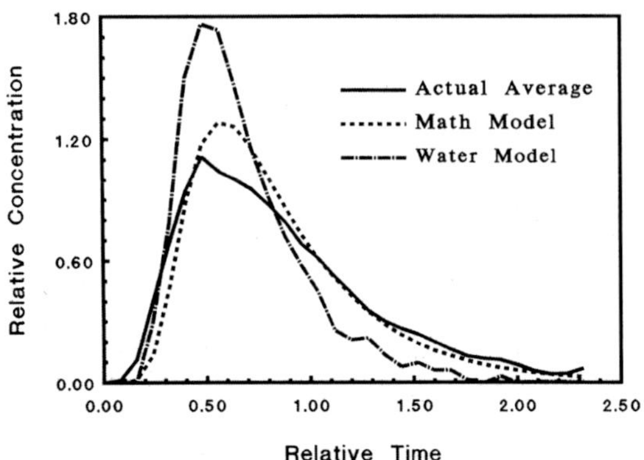

Figure 5.21: Comparison of actual and model residence time distributions for the inside nozzle of the tundish with the baffles installed. [Ref. 11]

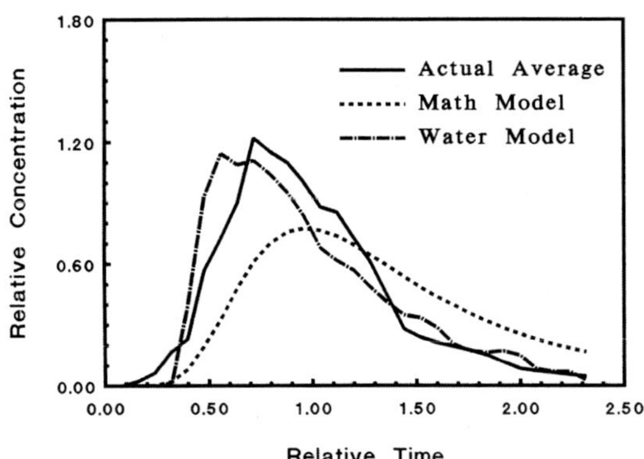

Figure 5.22: Comparison of actual and model residence time distributions for the outside nozzle of the tundish with the baffles installed. [Ref. 11]

5.4.3 *Case study # 3*

Mukhopadhyay *et al.* [13] studied the effects of different flow control devices in tundishes on flow and turbulence profiles and finally on inclusion flotation. They mathematically modeled flow in a tundish

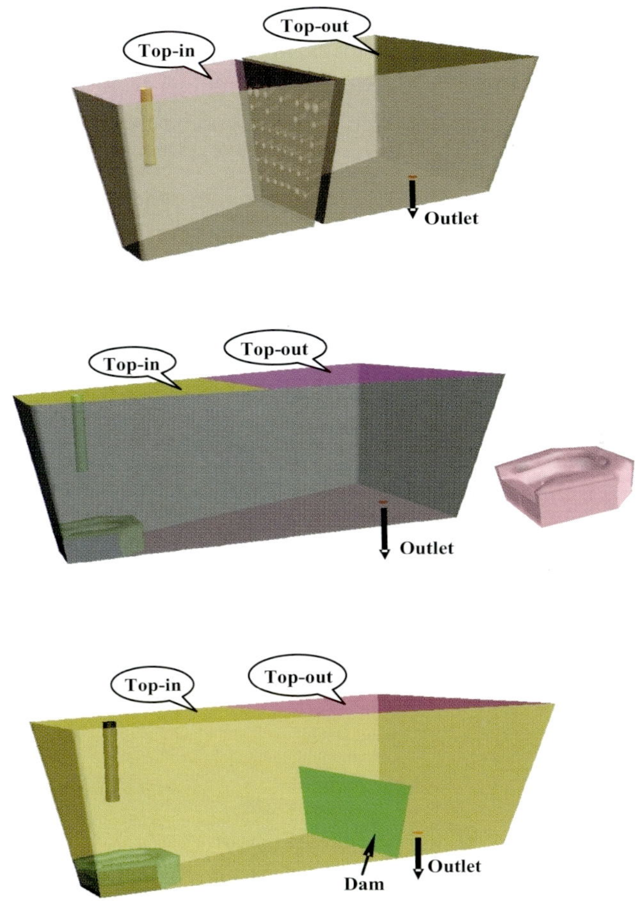

Figure 5.23: Tundish with a baffle with holes (top), a pour pad (middle), and with a pour pad and dam (bottom). [Ref. 13]

with a baffle with holes, a pour pad, or a pour pad with a dam. These three tundish configurations are shown in Fig. 5.23. Top-in and top-out in Fig. 5.23 and later in Fig. 5.26 refer to inclusions floated to the top surface on the inside and outside, i.e. upstream and downstream of the baffle location. Velocity vectors colored by velocity magnitude and turbulent kinetic energy iso-surfaces in the tundish with a baffle are shown in Figs. 5.24 and 5.25. Similar predictions were also made with the other two configurations. The authors studied the flotation of 10 to 500 micron sized inclusions.

Fig. 5.26 shows the percentage of inclusions which floated to the top surface and escaped through the outlet to the cast steel. It can be seen that inclusions greater than 100 microns were completely trapped at the top surface and did not exit through the tundish outlet. Further analysis of the data in Fig. 5.26 was carried out to show the statistical bounds of the residence time data, which is shown in Fig. 5.27. It shows that very small inclusions are subject to maximum influence of fluid turbulence and that the residence time varied from 1200 seconds to 150 seconds. The residence time variation decreases with increasing particle size, and becomes indistinguishable at sizes greater than 200 microns. The unsteady behavior of inclusions in a turbulent metal bath was very well simulated in this study. Thus, mathematical modeling using CFD can be very effectively used for tundish design analysis and optimization.

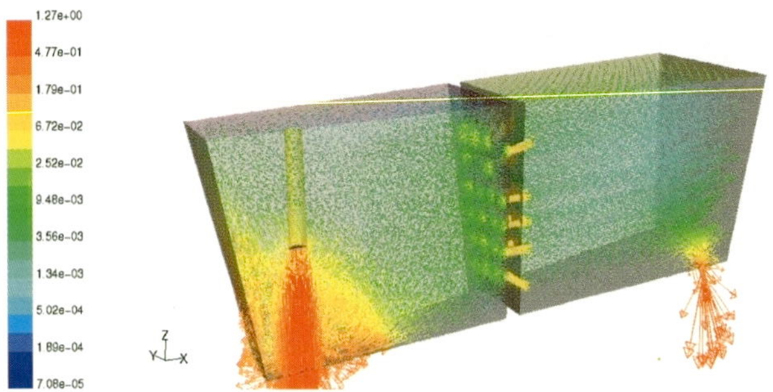

Figure 5.24: Velocity vectors colored by velocity magnitude in the tundish with baffle. [Ref. 13]

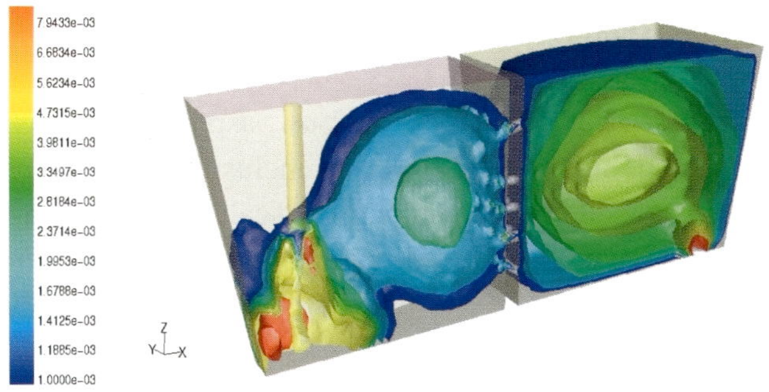

Figure 5.25: Turbulent kinetic energy iso-surfaces in the tundish with baffle [Ref. 13]

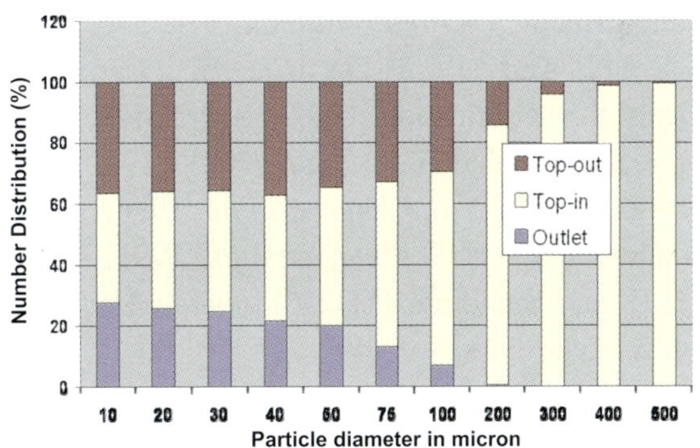

Figure 5.26: Distribution of inclusion particles on the top surface and in the outlet stream. [Ref. 13]

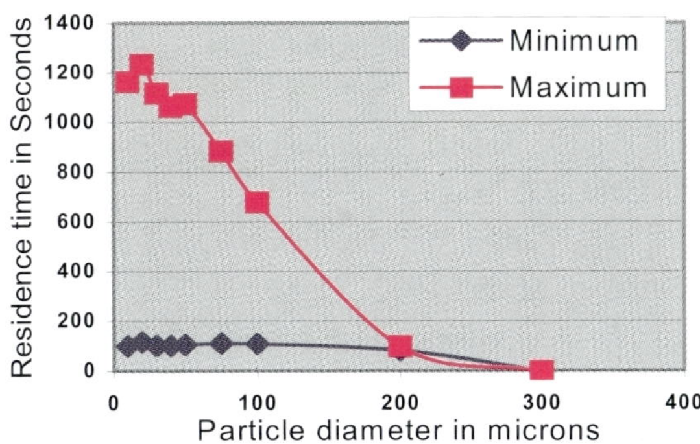

Figure 5.27: Distribution of particle residence time with a baffle. [Ref. 13]

5.5 Concluding Remarks

This chapter has covered various aspects of physical and mathematical modeling of melt flow and inclusion removal in tundishes. Various factors considered in deciding the scale for a tundish model and similarity criteria were discussed in detail. A brief overview of the mathematical modeling of melt flow was presented. Finally, selected case studies of water and physical modeling and tundish design were presented.

References

1. Sahai, Y. and Emi, T.: *ISIJ International*, 1996, 36, 1166-1173.
2. Sahai, Y. and Burval, M.D.: *Electric Furnace Conference Proceedings*, I.S.S. Publication, 1992, 50, 469-474.
3. Chakraborty, S. and Sahai, Y.: *Metall. Trans.*, 1992, 23B, 152-167.
4. Damle, C. and Sahai, Y.: *ISIJ International*, 1996, 36, 681-689.
5. Damle, C. and Sahai, Y.: *ISIJ International*, 1995, 35, 163-169.
6. Launder, B.E. and Spalding, D.B.: *Computer Methods Appl. Mech. Eng.*, 1974, 3, 269-289.
7. Chakraborty, S. and Sahai, Y.: *Sixth International Iron and Steel Congress*, ISIJ Publications, 1990, 3, 189-196.
8. Patankar, S.V.: *Numerical Heat Transfer and Fluid Flow*, Hemisphere Pub. Corp., New York, NY, 1980.
9. Sinha, A.K. and Sahai, Y.: *ISIJ International*, 1993, 33, 556-566.
10. Harris, D.J. and Young, J.D.: *Proc. Steelmaking Conference*, ISS Publication, 1982, 65, 3-16.
11. Lowry, M.L. and Sahai, Y.: *Steelmaking Proceedings*, I.S.S. Publication, 1989, 72, 71-79.
12. Lowry, M. L. and Sahai, Y.: *Transactions of the ISS*, published in Iron & Steelmaker, March 1992, 81-86.
13. Mukhopadhyay, A., Gilles, H.L., and Kocatulum, B.: *Electric Furnace Conference*, I.S.S. Publication, 2002, 60, 343-351.

Chapter 6

Tundish Operation

6.1 Introduction

One of the most important functions of a tundish, as mentioned in Chapter 1, is to continuously deliver molten metal of desired composition, temperature, and cleanliness from the ladle to one or more molds at the desired flow rate. If proper care is not taken during tundish operation, the improvements in metal quality brought about by various ladle refining operations may be totally negated during transfer of metal from the ladle to the tundish and during metal flow through the tundish to the molds. These transfers provide ample opportunity for the metal to interact with air, tundish slag, and refractories.

As discussed in Chapter 2, these interactions may generate additional macroscopic non-metallic inclusions in steel, and deteriorate its quality if the size of the macro inclusions exceeds critical values. Inclusions in a tundish can be broadly classified as exogenous and indigenous inclusions. The harmful macro inclusions originate mostly from the exogenous inclusions, and some form through the agglomeration of indigenous ones. Exogenous inclusions arise from reoxidation by air and oxidizing ladle slag, the entrainment of ladle slag and tundish flux, and the erosion of the refractory lining of the ladle or the tundish. Macro inclusions also occur from reoxidation during metal transfer from the tundish to the mold, by the entrainment of mold slag, and by the dislodging of alumina accretion from the SEN.

The melt remains in the tundish for a relatively short time, reflecting the continuous nature of tundish operation. Thus, the major refining reactions such as deoxidation and desulfurization are carried out in the ladle. Accordingly, tundish operation is designed to eliminate or

173

minimize the sources of macro inclusion formation and to help in the removal of macro inclusions from the melt, and these are described in this chapter.

6.2 Reoxidation During Ladle to Tundish Melt Transfer

The phenomenon of melt stream reoxidation during teeming from the ladle to the tundish and the benefits of eliminating this reoxidation are well documented in the literature [*e.g.* refs. 1-6]. The pre-eminence of the reoxidation phenomenon in the generation of large-sized inclusions can be seen from Fig. 6.1. Ohno *et al.* [4] found that metal stream reoxidation increased the amount of large inclusions by a factor of 2.5 between the ladle and the tundish, and that the reoxidation products were bigger than 100 microns in size. The teeming stream can be protected from the oxidizing atmosphere by physical shrouding where the melt stream is enclosed in a refractory pouring tube or in an envelope of Ar gas. Frequently, a combination of both of these methods is utilized. Fig. 6.2 [5] schematically shows argon gas shrouding inside of a refractory tube and the use of long-nozzle melt stream shrouding techniques.

The benefits resulting from melt stream protection are manifold. Demasi and Hartmann [6] observed a reduction in the average oxygen content of the bath and a consequent reduction in the number of alumina inclusions in Al-killed steel melt with ladle stream shrouding. A definite improvement in the surface quality of slabs was also observed when air reoxidation of the melt was prevented. The use of inert gas shrouding also led to an increase in the performance efficiency of the mold slag, as it would have fewer alumina inclusions to absorb. The low alumina mold slag, being more fluid, provided more consistent and uniform heat removal to the mold, which in turn resulted in the reduction of the slab surface cracks and reduced the incidence of breakout phenomenon. The reduction in the number of alumina inclusions also resulted in the decrease in clogging frequency of the tundish nozzle. Melt stream shrouding caused a reduction in the nitrogen pick-up by the melt during casting. Another interesting observation was that the use of shrouding resulted in the reduction of melt temperature loss from the ladle to the tundish by as much as 5 to 8 °C, allowing casting of melt in ladle at lower temperatures.

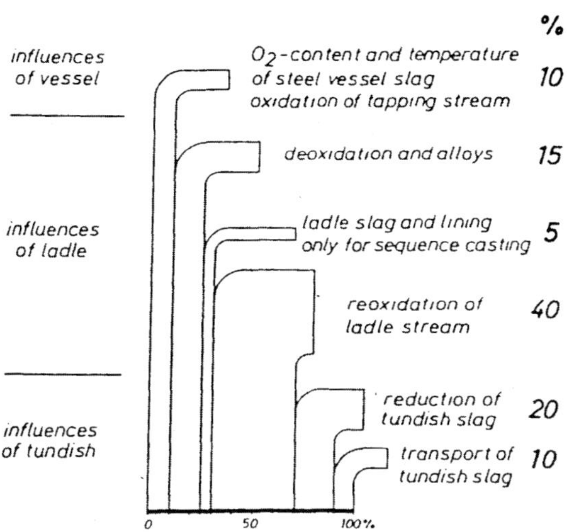

Figure 6.1: Relative contribution of various factors causing inclusions in steel. [Ref. 4]

The melt stream shrouding also has some shortcomings. If ladle slag is carried over in the inert gas shrouding setting, the impinging melt stream from the ladle emulsifies slag into the melt and forms macro inclusions. If the operating melt bath depth in the tundish is inadequate, the use of a long nozzle increases the impact pad erosion rate, which may result in an increase in large complex oxide inclusions. Turbulent melt flow caused by the initial period of long nozzle immersion during the ladle change also contributes to the occurrence of macro inclusions. The tundish flow with a long nozzle has been observed to be more stratified in some cases. Use of the submerged long nozzle also reduces the filler sand removal and prevents the visual identification of ladle slag carryover to the tundish. However, the danger of filler sand turning into macro inclusions has been somewhat reduced by modifying the constitution of the sand. The use of present-day slag sensing devices has significantly improved the inspection of slag carryover from the ladle to the tundish, and the need for visual observation has become less important.

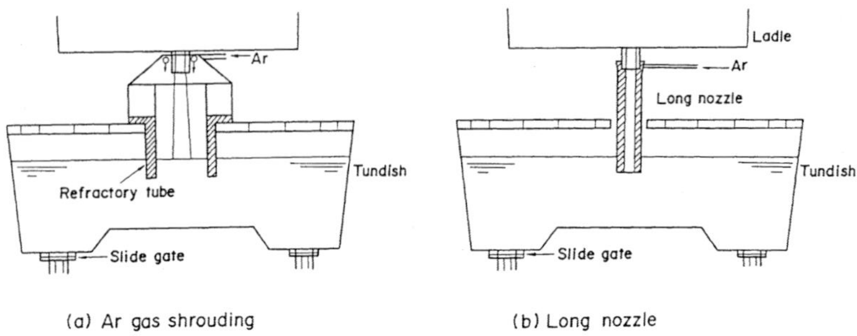

Figure 6.2: Ladle stream protection by (a) Ar gas shrouding and (b) long nozzle. [Ref. 5]

6.3 Slag Entrainment and Sensing Technologies

The oxidizing components, FeO, MnO, and SiO_2, in ladle slag carried over to the tundish react with aluminum in the steel and form alumina clusters in steel slabs. Part of the slag transferred from the ladle to the mold may get trapped in the solidifying shell, leading to the formation of macro inclusions or slag spots. Several studies [*e.g.* refs. 3, 8-10] have clearly demonstrated the importance of bath depth in the tundish, especially during ladle change, on the rate of slag entrainment by the melt flowing to the mold, as discussed in section 6.4. Vortexing may occur at low melt depth during tundish draining, which may carry a part of the tundish slag through the tundish outlet to the mold. Similar conclusions were reached by Cramb and Byrne [11] from their water model studies and actual plant trials. They added CeO_2 as a tracer to the tundish slag, and the mold slag was analyzed for this tracer to detect tundish slag entrainment. Figure 6.3 represents the results of one such tracer trial. The figure shows that the entrainment phenomenon occurs during the ladle change-over period. Turbulent conditions near the slag/metal interface during various operating conditions were identified as the major reason for slag entrainment. Full immersion of the ladle nozzle into the tundish shroud may not be possible during the filling operation of a tundish with a new ladle, subsequent to the ladle change operation.

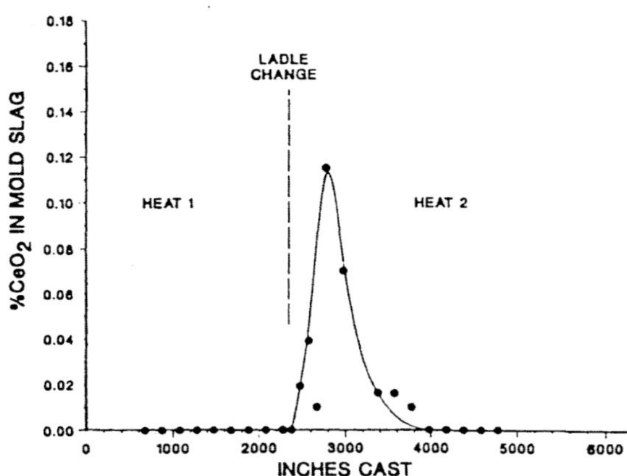

Figure 6.3: Tracer concentration in mold slag transferred from ladle slag during ladle change between two heats. [Ref. 11]

Slag detection devices can be used at the ladle-to-tundish outlet that can detect the onset of slag transfer from the ladle to the tundish. The following techniques have been used for the detection and minimization of slag carryover:

(1) Electromagnetic methods
(2) Visual observation
(3) Optical detection method
(4) Vibrational technique
(5) Weight monitoring technique
(6) Slag float valve

The electromagnetic method in which the slag sensor electromagnetically detects slag in the pouring streams is very efficient and is commonly employed in the steel industry. This was first developed by a company named AMEPA [12], which stands for Angewandte Messtechnik und Prozessautomatisierung, *i.e.* Applied

Instrumentation and Process Control. The AMEPA slag detection sensor is now being used in many countries.

The sensor consists of two coils, as shown in Fig. 6.4, and is placed around the nozzle at the bottom of the ladle. When a current is passed through the primary coil, it induces an electromagnetic field in the steel flowing through the nozzle. The field, in turn, generates eddy currents in the secondary coil. The electrical conductivity of slag is significantly lower than that of molten steel, and thus, the induced eddy currents with slag in the stream are lower than those with the metal. The eddy currents are continuously monitored, and the sensor distinguishes between the flow of slag and metal in the stream. The arrangement of the sensor, measurement and control equipment, and the recorder is schematically shown in Fig. 6.5.

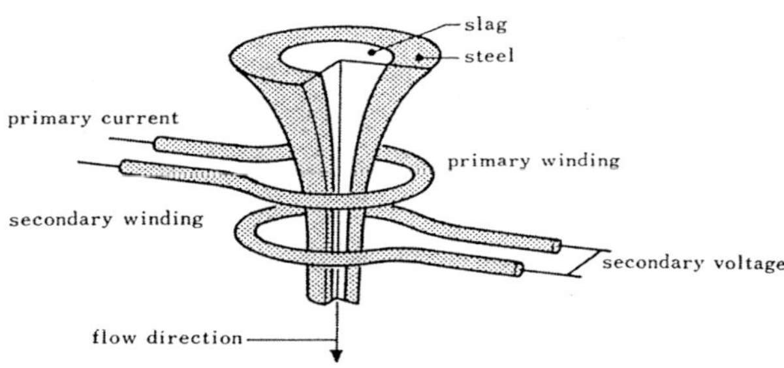

Figure 6.4: Primary and secondary coils of an electromagnetic sensor.

The sensor signal can directly operate the slide gate nozzle, and thus, can minimize slag carryover. Figure 6.6 shows the slag signal as detected by the AMEPA sensor [13]. The figure demonstrates that it takes less than 2 seconds for the sensor to activate the closing of the slide gate nozzle aperture. Thus, the number of defects in slabs can be significantly reduced as shown in Fig. 6.7 [14]. The use of this sensor has resulted in a substantial reduction in carried-over ladle slag to the tundish, even after several ladles in a sequence casting, which in turn reduces the defects in cast slabs.

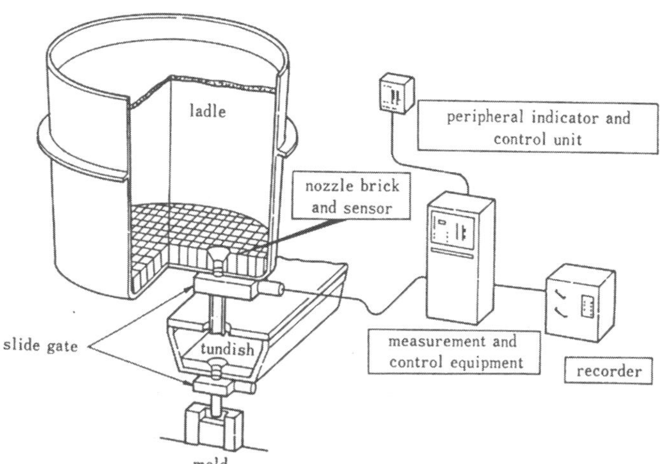

Figure 6.5: Electromagnetic sensor measurement and control equipment, and recorder arrangement.

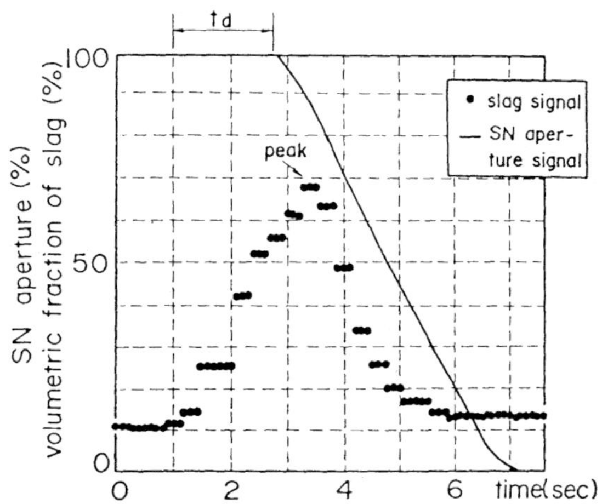

Figure 6.6: Slag carryover signal detected by AMEPA sensor. [Ref. 13]

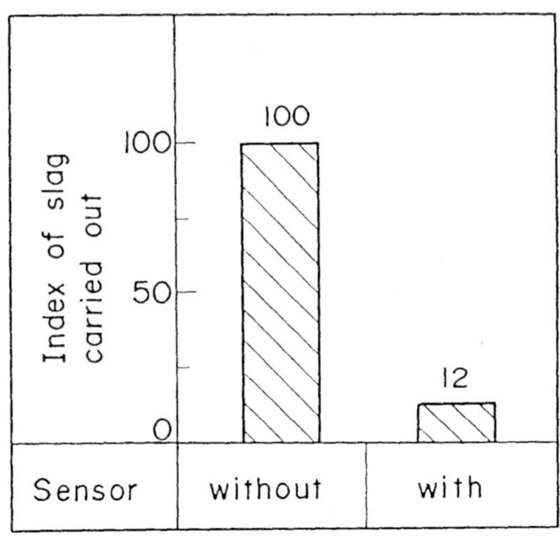

Figure 6.7: Reduction in slag index by the use of AMEPA sensor. [Ref. 14]

Most of the other methods listed above were more commonly used before the development of the electromagnetic sensor. As the name suggests, the *visual observation method* relies on the operator's capability to spot the inflow of slag, and thus, may result in significant slag carryover before stopping the melt flow. The principle of slag detection by an *optical sensor* is based on the difference in emissivity and stream diameter between the slag and steel. Both the emissivity and the stream diameter are greater for slag than for steel. Reference [15] discusses details of this system and presents some of its results. It is reported that in 80% of cases, the optical system warning alarm appeared 5 seconds before visual observation of the slag. The system does not work satisfactorily when a vortexing slag core is surrounded by the metal stream, or when the slag stream is hidden on the opposite side of the metal stream monitored by the sensor.

Vibrational analysis sensors monitor the vibrations generated on the ladle to tundish nozzle. These vibrations are inherently different for the flow of steel and steel mixed with slag. This method is, however, sensitive to the background vibrations. Itoh *et al.* [16] used such a sensor on a long nozzle between the ladle and the tundish. The signal from the vibrational sensor activated the slide gate nozzle. The authors of the study found that detection by the sensor was at least 3 seconds quicker than the visual response.

The *weight monitoring technique* relies on the difference in the steel and slag densities. This technique analyzes the change of ladle weight with time. The rate of change in the teeming rate is continuously monitored, and any rapid change in this rate relates to the change from the metal flow to the mixed metal and slag stream. References [17-18] discuss the method and give some of the metallurgical results.

Slag float valve in the ladle - Vortexing may start in the final stages of any sequence casting at lower melt depths, and ladle slag may transfer to the tundish. To prevent this transfer, a simple device called a *float valve* has been found to be very effective. The refractory material of the valve has a density greater than molten slag and less than that of steel. Thus it floats at the slag-metal interface. The device is schematically shown in Fig. 6.8. The valve helps to prevent drainage of the ladle slag into the tundish. Matsumoto *et al.* [19] analyzed the slab samples cast during a steady and non-steady state, with and without the float valve at Nippon Steel Corporation's Hirohata Works. The results are presented in Fig. 6.9, which demonstrate that the use of a valve is effective in reducing the macro inclusions in both steady state and non-steady state casting.

6.4 The Effect of Tundish Size

Tundish size has been found, as is partly discussed in Chapter 2, to have a significant effect in improving the quality of cast steel. Increasing the tundish size is the most obvious way of increasing the average residence time of steel in the tundish. Over the years, tundish size has gradually increased. Figure 6.10 [Ref. 20] shows that the trend is toward increasing tundish capacities. Table 6.1 lists several large capacity tundishes operating in Japan.

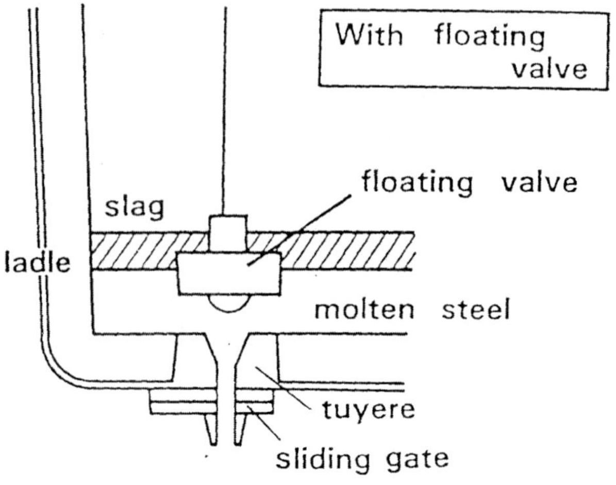

Figure 6.8: Float valve for prevention of ladle slag transfer. [Ref. 19]

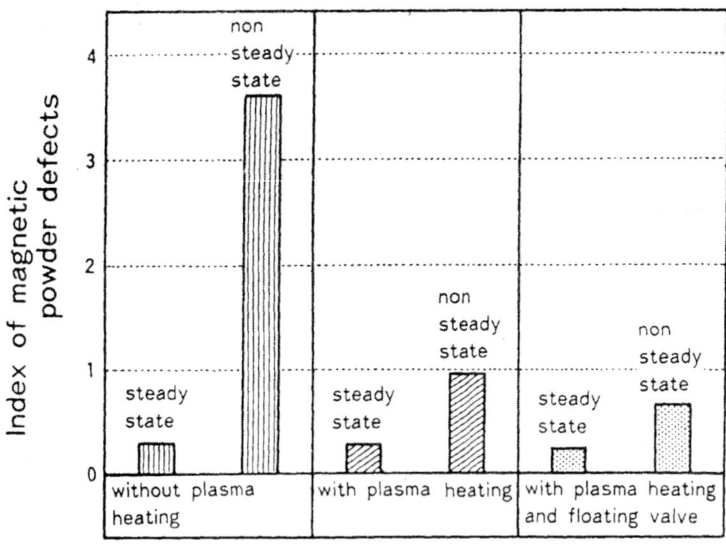

Figure 6.9: Effectiveness of float valve in reducing macro inclusions. [Ref. 19]

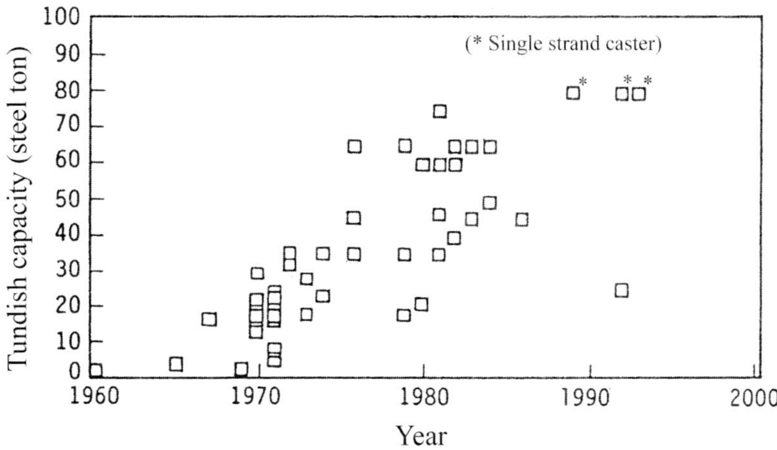

Figure 6.10: Increasing tundish capacity with years. [Ref. 20]

Table 6.1: Large capacity tundishes in Japan.

Continuous Caster	Tundish Capacity (tons)	Note
Nippon Steel Nagoya No. 2	60	H-Shaped Tundish
Kimitsu Nos. 2 and 3	60	
Ohita Nos. 4 and 5	70	
JFE (NKK) - Fukuyama No. 5	80	
JFE (Kawasaki) - Chiba No. 3	75	
JFE (Kawasaki) - Mizushima No. 4	70	
Sumitomo - Kashima No. 3	65	Increased to 85 tons
Kobe - Kakogawa No. 4	80	
Nisshin - Kure No. 2	65	

The beneficial effects of operating with a large tundish have been established by both water modeling and mathematical modeling studies, and by actual plant trials as mentioned in Chapter 2 [also see refs. 3, 8-10, 21, and 22]. For a given casting rate, which determines the volumetric flow rate of the fluid through the tundish, a larger volume results in a longer average residence time in the tundish. Increasing the volume of the tundish leads to a dramatic decrease in the number of macro inclusions (alumina clusters) obtained in the cast product particularly at ladle change, as seen from Fig. 6.11 [8].

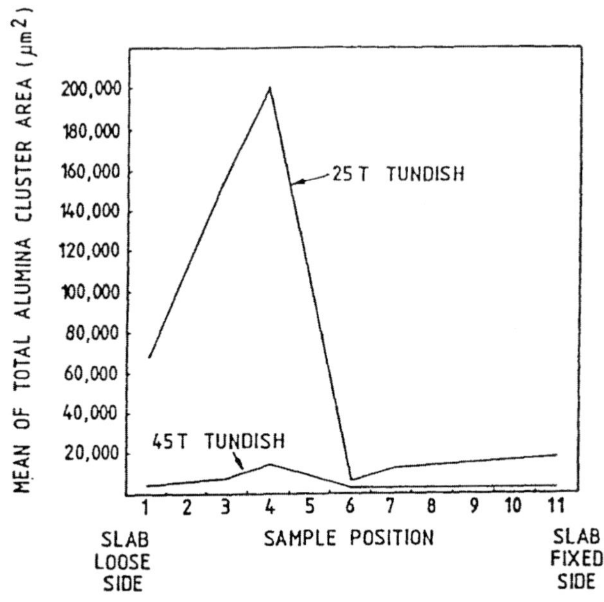

Figure 6.11: Effect of tundish size on alumina inclusions. [Ref. 8]

The volume of a tundish may be increased by increasing its length, width, or height. It is interesting to speculate on the relative importance of individual dimensions in determining the separation efficiency of an optimally designed tundish. Tacke and Ludwig [21], in a mathematical modeling study, investigated the relative importance of tundish width and bath depth in determining the removal rate of inclusions. In their study, the original tundish length was 3500 mm, while the depth and width were kept fixed at 700 mm each. The removal rate of the inclusions based on a simple Stokian flotation model was calculated to be 41.3 %. The volume of the tundish was then increased by increasing the width to 1200 mm and keeping the other dimensions the same. The removal efficiency increased to 55.4 %. In contrast, by increasing the depth to 1200 mm, while keeping the other dimensions the same, the removal rate dropped to 38.5 %. This is because increasing the volume by increasing the bath depth affected two factors. First, increasing the tundish volume increases the fluid residence time in the tundish. Second, increasing the depth increases the distance for a particle to rise to the surface. These two factors therefore have opposing effects on inclusion flotation to the free surface.

In reality, the situation is somewhat different. Removal of the inclusion particles by allowing them to rise to the surface by Stokes' velocity remains the most influential factor. However, as discussed in Chapter 2, removal is also influenced by factors other than the Stokes' flotation. Factors contributing to the removal and formation of inclusions include:

(1) Coalescence of particles by turbulent collision in the inlet section and by gradient collision in the outlet section of a tundish;

(2) 3-D melt flow modification caused by a thermal gradient in the outlet section of the tundish, where the upward component of the flow helps the Stokes' flotation directly. The flotation also enhances the gradient collision in forming larger inclusion clusters which rise faster by the Stokes' flotation;

(3) Sticking of the particles to refractory surfaces of the tundish;

(4) The effect of overall flow pattern in the tundish on the entrainment of and reoxidation by tundish slag; and

(5) Reoxidation by air across the tundish slag and by oxidizing components in the tundish slag.

Sinha [22] mathematically modeled the inclusion removal rates by considering flotation, sticking to the refractory, and the effect of coalescence due to turbulence. He studied the effect of varying length to width (L/W) and height to width (H/W) ratios of a typical slab caster tundish on the inclusion removal rates. He found that the use of a shallow tundish (low H/W) resulted in excellent inlcusion removal rates but shallow tundishes may be limited by vortexing phenomenon at low bath depths. The removal rates decreased with the increase of bath depth until an H/W ratio of 1, beyond which they increased until an H/W ratio of 1.2

to 1.3. The removal rates deteriorated again with the increase of an H/W ratio beyond 1.3.

The results of an actual plant trial are shown in Fig. 6.12 [3], in which the effect of the bath depth on inclusion number was studied. It shows that the inclusion content decreases as the bath depth is increased. The same publication [3] compared the degree of reduction of defects arising from the slag and from alumina clusters with the increase of bath depth, shown in Fig. 6.13. The results indicated a dramatic decrease in the incidence of slag defects but some increase in the number of alumina inclusions when the bath depth was increased.

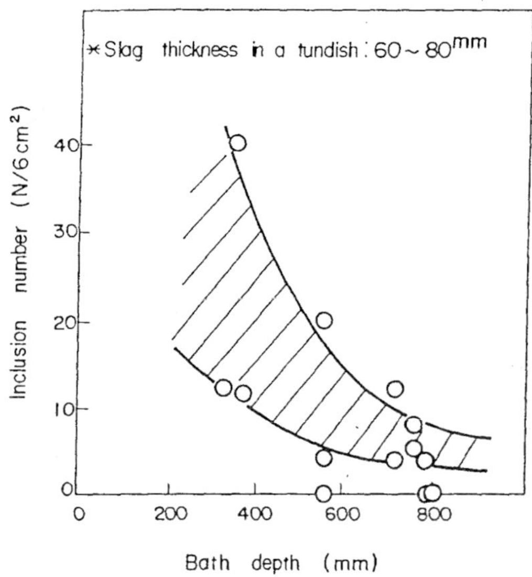

Figure 6.12: Effect of bath depth on inclusion number. [Ref. 3]

The results of another plant trial with a much larger tundish by Tozaki *et al.* [9], where the tundish capacity was increased from 65 to 85 tons, are shown in Figs. 6.14 and 6.15. The number of alumina clusters decreased during both the steady state and non-steady state operations. Another benefit of increasing the depth of the tundish was in maintaining the casting rate without any decrease during ladle change. Thus, the productivity and quality both were improved by increasing the depth of the tundish. Similar plant trial results made at Kobe steel were published by Ishikura *et al.* [10]. In this study, the performance of an old caster (No. 3) with a 50-ton tundish is compared with their new caster (No. 4) with an 80-ton tundish. They found that the caster with the 80-ton tundish can operate at 2.0 m/min casting speed and cast the same quality of slabs as the caster with the 50-ton tundish at 1.4 m/min speed (see Fig. 6.16). Thus, the productivity of the caster was increased without sacrificing the quality. They also found that the quality of the slabs cast during the ladle change period was significantly improved with the 80-ton tundish caster, as shown in Fig. 6.17. The normal operating depth of the 80-ton tundish was 2000 mm, which was achieved by making the bottom of tundish inclined toward the outlet. These studies clearly demonstrate the benefit of greater tundish depth and size on the quality of cast slabs.

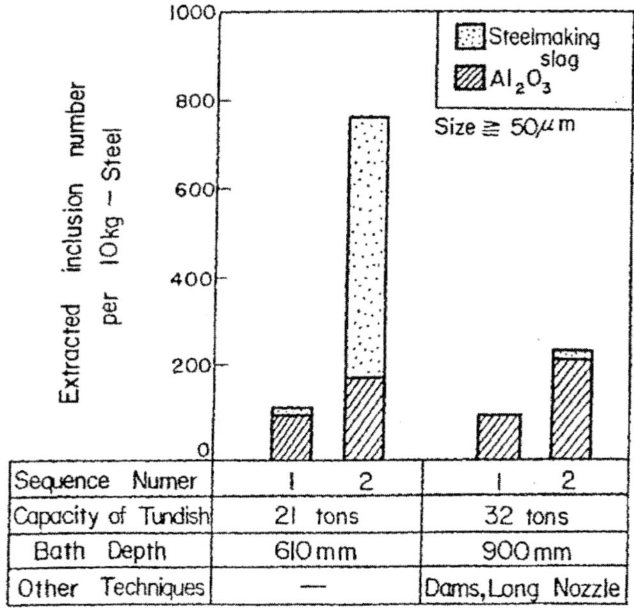

Sequence Numer	1	2	1	2
Capacity of Tundish	21 tons		32 tons	
Bath Depth	610 mm		900 mm	
Other Techniques	—		Dams, Long Nozzle	

Figure 6.13: Effect of tundish bath depth on inclusion number. [Ref. 3]

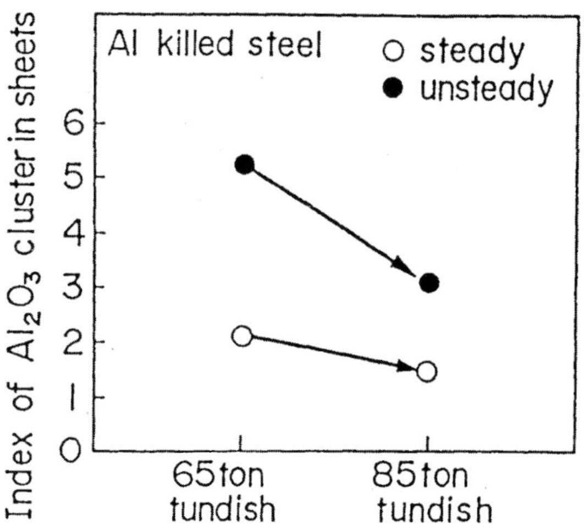

Figure 6.14: Effect of bath depth on alumina clusters. [Ref. 9]

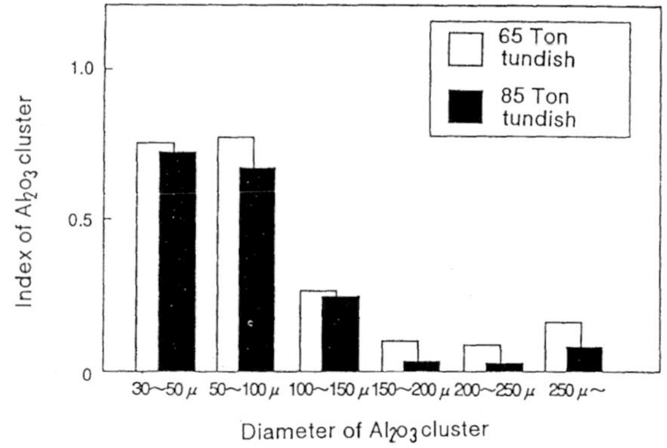

Figure 6.15: Effect of bath depth on alumina clusters. [Ref. 9]

6.5 The Effect of Flow Control Devices

Many attempts have been made to improve melt flow characteristics in existing tundishes by the installation of various flow control devices, such as *weirs, dams, baffles with holes, pour pads, and turbulence suppressers,* within the tundish. The beneficial effects of various flow modification devices have been borne out by actual industrial trials as well as physical and mathematical modeling studies [*e.g.* refs. 21-30]. Optimum placement of dams and weirs has been found to result in an increase in the average residence time of fluid as well as an increase in the plug flow volume in the tundish.

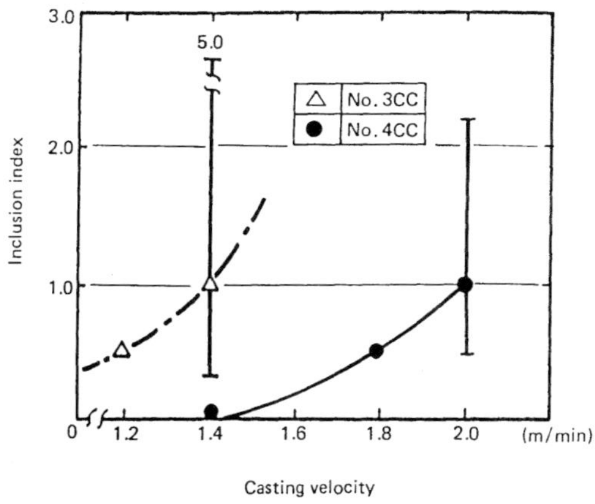

Figure 6.16: Effect of casting speed on inclusion index. [Ref. 10]

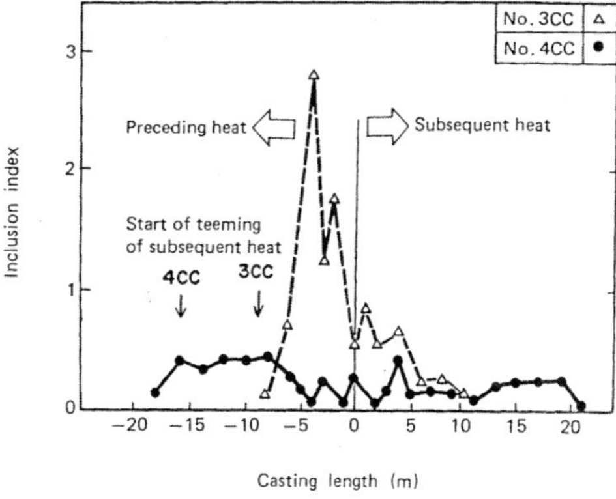

Figure 6.17: Quality improvement in 80 ton caster. [Ref. 10]

The studies of Kemeny *et al.* [23] and Young and Harris [24] have shown the benefits resulting from the use of flow modification devices in their tundish. Three industrial case studies presented in section 5.4 describe the benefits achieved by the use of tundish flow control devices in improving cast steel quality. Extensive water modeling studies of Inland Steel's No. 3 caster tundish were carried out to find the optimum number and placement of weir/dam pairs [25]. It was found that an arrangement of weir/dam/weir/dam on either side of the ladle shroud produced the maximum volume fraction of plug flow and maintained the best strand similarity. In their study, the heights of the dams and weirs used were found to have almost no effect on the extent of plug flow or strand similarity. Similarly, the presence of stopper rods and the submergence of the ladle shroud were found not to have any significant effect on tundish performance, implying the pre-eminence of the dams and weirs in influencing the fluid flow profiles in the tundish.

The third type of flow control device commonly used in an industrial tundish is a *baffle with holes*. Figure 6.18 is a schematic diagram of the baffle that was used at the Sparrows Point caster of Bethlehem Steel Corporation [26]. In their practice, broken dams and weirs resulted in operational difficulties; hence, it was decided to replace the dams and weirs by a single baffle with holes. The flow pattern generated by the baffle was found to be similar to that produced by the dam/weir arrangement. Use of a multiple hole baffle in a 6-strand tundish in the Armco Kansas City Works resulted in a more uniform distribution of steel to the casting nozzles, an increase in the mean residence times for nozzles closest to the ladle shroud, and increased melt mixing [27]. Dorricott *et al.* [28] suggested use of a baffle with asymmetrically arranged holes in a multistrand tundish for improved fluid flow to all strands. The average surface slag index of the cast product was found to decrease with the use of an asymmetric baffle.

These flow control devices, properly installed, may create localized mixing in contained regions, which may help in inclusion agglomeration and hence, their removal. In a laboratory test, Emi and Habu [29] at Kawasaki Steel used different configurations of flow control devices, as shown schematically in Fig. 6.19. They determined the Peclet number (Pe) for each flow design, and found that a smaller Pe value represented higher mixing intensity in the tundish. These flow configurations were also used in actual operation of a small size tundish (30-ton). The resulting defect index is plotted in Fig. 6.19. It is obvious that the configuration C with the smallest Pe value, resulting from the cross flow between the two baffles, was most efficient in inclusion removal from the melt.

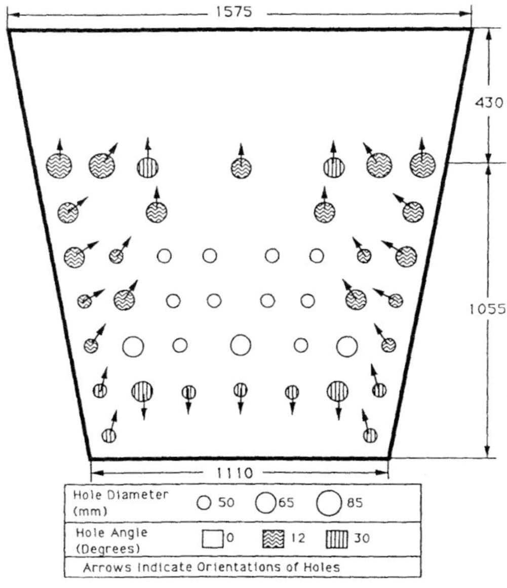

Figure 6.18: Schematic of a baffle with holes. [Ref. 26]

In spite of the overwhelming industrial evidence in favor of the effectiveness of the flow control devices in removing inclusions from the melt, some of the modeling studies [*e.g.* refs. 21, 22, 30], either physical or mathematical, have failed to show their effectiveness in inclusion removal. Tacke and Ludwig [21], and Sinha [22], in their mathematical modeling studies, showed that a dam and weir configuration did not lead to an increase in inclusion removal efficiency compared to the no-flow control case. Nakajima *et al.* [30], in their water modeling study employing hollow glass microspheres to simulate inclusions, actually reported a decrease in the inclusion removal efficiency with the incorporation of dams and weirs. It has to be kept in mind that any improvement in fluid flow characteristics in the tundish with the incorporation of flow control devices is dependent on the optimum location and size of the dams and weirs. Tolve *et al.* [31] have shown that the retention time of the fluid in a multistrand tundish is directly dependent on the distance of the dams from the ladle stream (see Fig. 6.20). Thus, an inappropriate placement of the flow control devices may result in a flow profile which is more detrimental, compared to the no-flow control case, as regards inclusion flotation and removal. The results of Nakajima *et al.*'s [30] study also point to the inadequacy of the glass microspheres in the simulation of the non-metallic inclusions in steel. Inclusion removal mechanisms, such as coalescence of the particles and sticking to the refractory walls, are very difficult to simulate with glass microspheres in a water model. The phenomenon of absorption of inclusions into the slag layer was also not simulated in Nakajima *et al.*'s study.

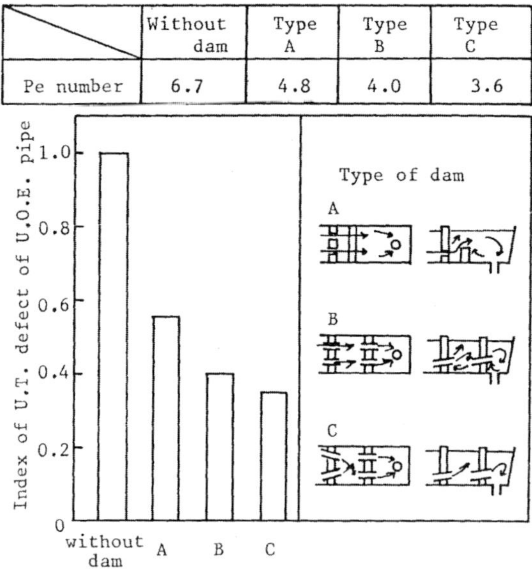

Figure 6.19: Effect of different flow configurations on defect index by ultrasonic test. [Ref. 29]

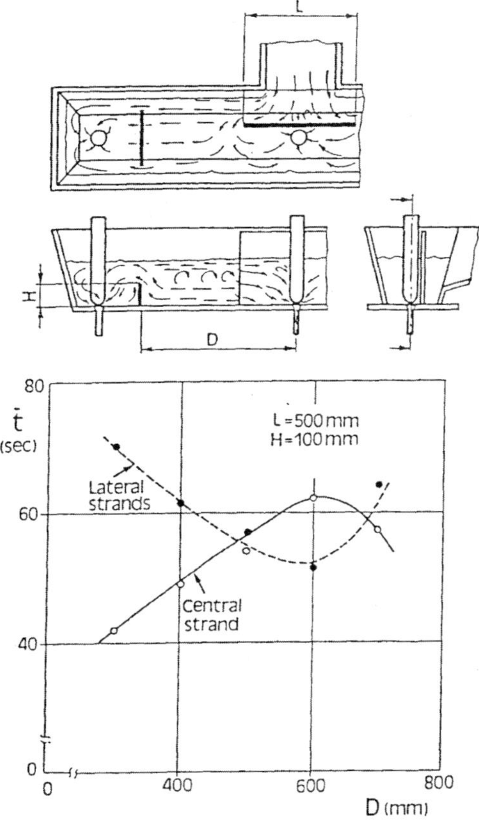

Figure 6.20: Effect of the dam distance on retention time. [Ref. 31]

Pour Pads and Turbulence Suppressers – The metal stream from the ladle, especially when it is shrouded and has no entrained gas, enters into the tundish at a very high velocity and turbulence. The impact of the stream at the tundish bottom may cause severe refractory erosion problems. Pour pads are designed and placed at the bottom to withstand the erosive force of the ladle stream. They are made of very dense, chemically stable refractory material, and are traditionally designed to be horizontally flat. Recently, several other shapes in which the pour pad had curvature or was inclined at some angle to divert the incoming stream in a desired direction, have been successfully tried [*e.g.* Fig. 6.21 from ref. 26]. Corrugated impact pads were found to be effective in reducing the turbulence of the incoming stream. A pour pad design which appears to reduce the stream turbulence significantly has been reported by Bolger and Saylor [32]. Their pour pad and associated fluid flow are shown in Fig. 6.22. The plunging metal stream, after diverging back to the free surface, becomes quiescent. Such surface directed flow is considered favorable for inclusion flotation. Of course, the surface directed flow depends upon the original shape of the pad which may change with refractory erosion.

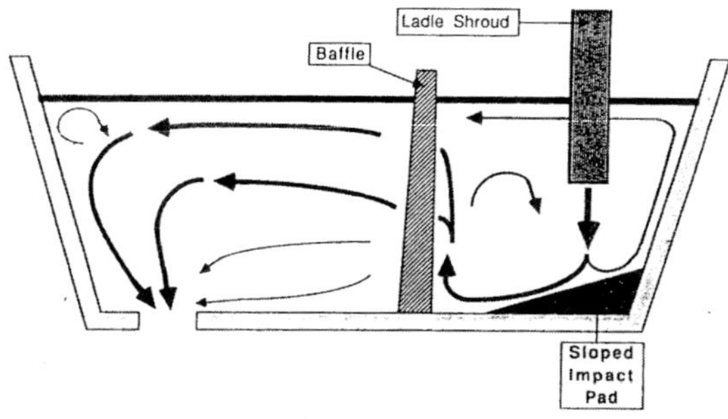

Figure 6.21: Schematic of a sloped impact pad in a tundish. [Ref. 26]

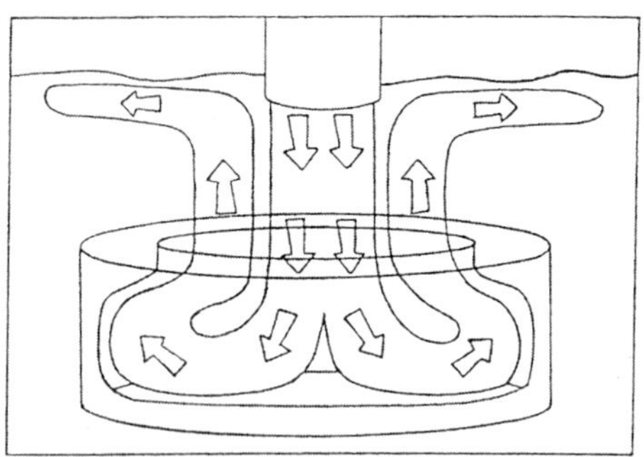

Figure 6.22: Schematic of a pour pad and associated melt flow patterns. [Ref. 32]

6.5.1 *Large tundish without flow modifiers*

The benefits of implementing various flow modifiers, mentioned above, on the cleanliness of melt have been changing recently. Significant developments in secondary refining processes and their installation in many steel plants have achieved acceptable levels of steel melt cleanliness for demanding applications, especially during steady state casting. Even with careful control of secondary refining, macro inclusions do occur during the non-steady state of casting. Large tundishes that offer sufficient residence time for macro inclusion flotation and that prevent vortexing at non-steady state have been found satisfactory for sequential casting of high quality steels, even at a relatively high casting rate. In addition, flow modifiers are inconvenient to sustain hot cycle tundish practice, which has been found to be quite economical. With a large tundish, the deep melt bath does not require a complicated pour pad, either. A large tundish of simple design, which can cast at a high throughput rate and sustain long sequential casting, would result in high productivity and high quality at low cost.

6.6 Gas Injection in Tundishes

Several water modeling studies and plant trials [*e.g.* refs. 33-36] have demonstrated the beneficial effects of Ar gas injection in a tundish. Yamanaka *et al.* [33] studied the effects of Ar gas injection in one arm of a V-shaped tundish. Double dams were installed in both arms of the tundish, but only one side was equipped with a porous plug for Ar gas injection as shown in Fig. 6.23. The results are shown in Figs. 6.24 to 6.26. The number of large inclusions was reduced (Fig. 6.24), while the number of small inclusions was found to increase (Fig. 6.25). The index of ultrasonic defects that represent the number of macro inclusions reduced significantly (Fig. 6.26).

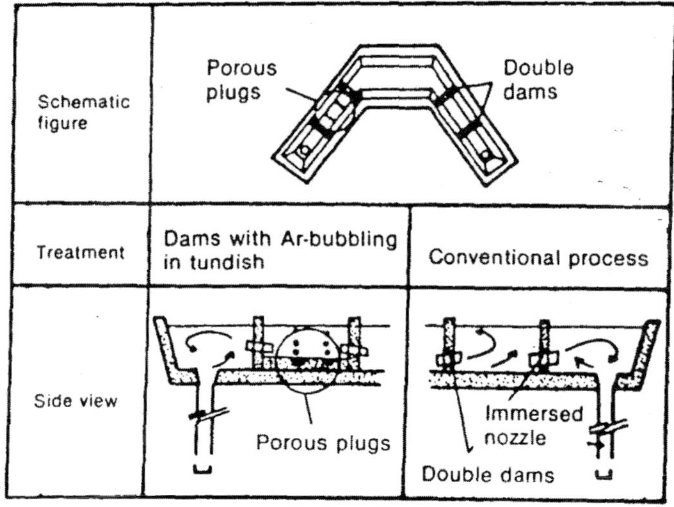

Figure 6.23: Gas injection in a V-shaped tundish. [Ref. 33]

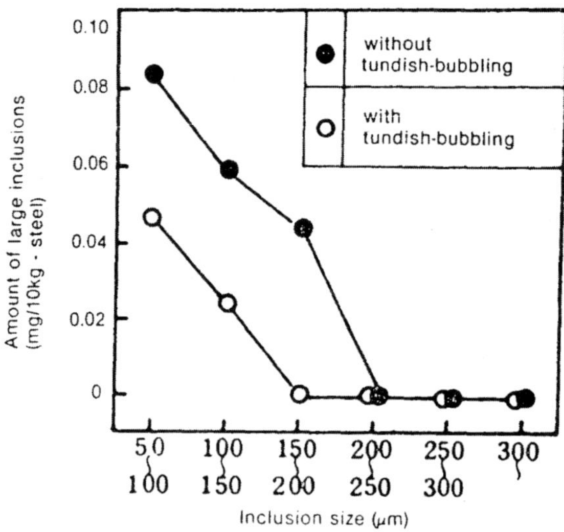

Figure 6.24: Effect of gas bubbling on large inclusions. [Ref. 32]

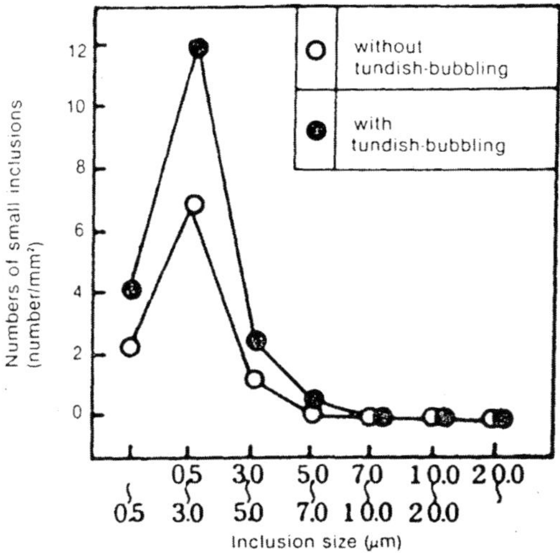

Figure 6.25: Effect of gas bubbling on small inclusions. [Ref. 32]

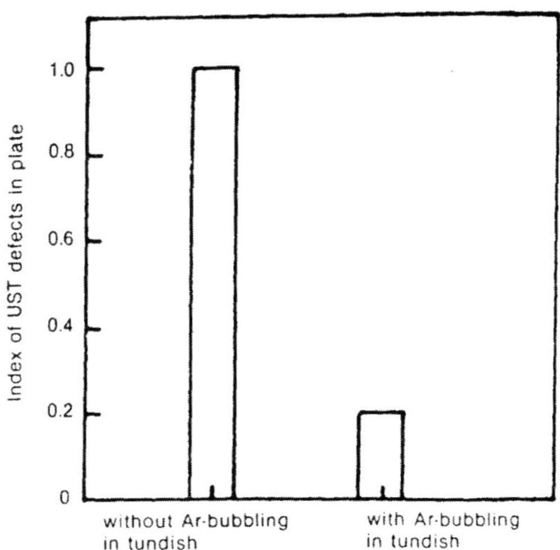

Figure 6.26: Effect of Ar gas bubbling on the plate defect index. [Ref. 32]

In another study at CRM [34], a gas bubbling device was embedded in the porous refractory at the bottom of a tundish (see Fig. 6.27). They also used a water model of the system to study the fluid flow pattern and its effect on simulated inclusion flotation. The fluid flow patterns, schematically shown in Fig. 6.28, suggest that the horizontal velocity of water in the gas injection region was reduced. Gas injection also increased the removal rate of simulated inclusions in the size range of 20 to 100 microns from 43% to 65%. Inclusions larger than 100 microns were being nearly completely removed even without gas injection. Positive metallurgical results in industrial trials showed that the cast product contained reduced amounts of both the hydrogen content and oxide inclusions. They found that the oxide area fraction, which is directly related to its inclusion contents, of steel treated by Ar gas injection was 25 to 50% lower than the untreated metal.

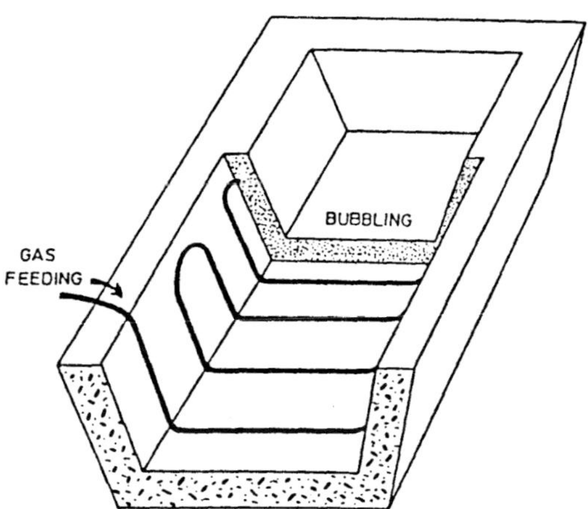

Figure 6.27: Ar gas bubbling device embedded in the porous refractory. [Ref. 34]

Gas in a tundish can be injected from a porous plug installed at the bottom of the tundish as well as from a lance inserted from the top. In a two-strand symmetrical caster, a rotating lance was inserted on one side of the inlet stream to generate fine gas bubbles [35]. The schematic side view of the rotating nozzle in a tundish is shown in Fig. 6.29. The effects of the Ar gas flow rate and the rotation speed on the inclusion removal were studied. The results obtained in this study, given in Fig. 6.30, clearly demonstrate the ability of the gas injected by the rotating nozzle to remove the smaller inclusions. The figure also compares the results of rotating the nozzle with bottom bubbling at the same gas flow rate but without rotation of the nozzle, and shows a definite improvement with the rotation of the nozzle. The rotating nozzle produces smaller bubbles and thus, is more efficient in inclusion flotation and removal. Figure 6.31 [36] shows the effects of Ar gas flow rate and speed of rotation on the inclusion removal efficiency. The results suggest that there is an optimum rotation speed and gas flow rate, 100 rpm and 2 to 6 l/min, respectively, for these experimental conditions, which produced the best results. It may be more instructive to study the effect of rotation speed and gas flow rate on the bubble size and liquid flow pattern to understand the role of these parameters on inclusion removal.

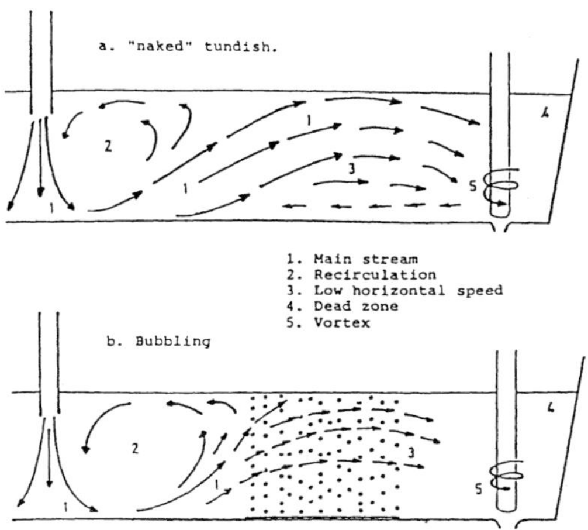

Figure 6.28: Effect of gas injection on melt flow patterns [Ref. 34]

In spite of the conclusive evidence of the beneficial effects of inert gas injection in melt contained in a tundish, gas injection is not commonly used in industrial practice. This may be attributed to gas injection's disadvantages. The gas injection removes the flux from the free surface and exposes metal to possible reoxidation. Even in a covered tundish with a protective atmosphere, it is not possible to eliminate completely the presence of small amounts of oxygen from the gas atmosphere. Thus, reoxidation of metal and thus, formation of oxide inclusions is always a possibility. Gas injection causes tundish slag entrainment at the slag/metal boundary due to turbulence arising from the swelling and break-up of gas bubbles. Gas injection also causes refractory erosion near the nozzle, which requires the use of more expensive refractory practices. The additional cost of refractory brings an economic factor into consideration. The third important factor is the reduction of melt temperature due to large volumes of gas injection, which in some cases may even freeze the metal and form skull. The gas injection practice should be coupled with external heating, and in fact, it is used with plasma heating for temperature homogenization. Thus, gas bubbling has shown metallurgical benefits, but is also associated with disadvantages. Its use may become more important if the tundish takes up some refining or compositional adjustment functions in the future, which may require mixing in some regions of the tundish.

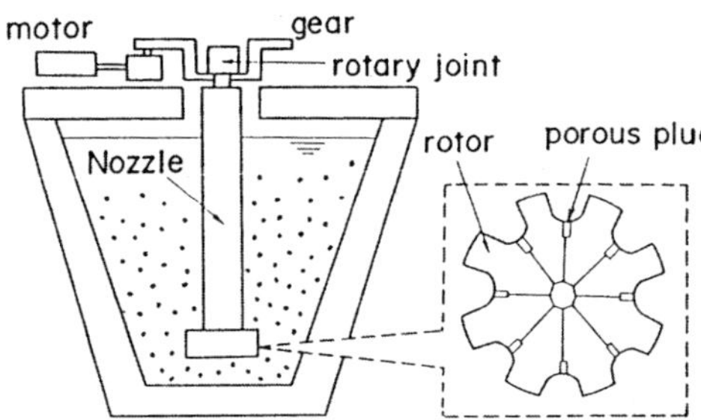

Figure 6.29: Schematic of a rotating nozzle in a tundish. [Ref. 35]

The current trend is, however, toward the opposite direction as mentioned in the previous section. Adjustment of the chemistry and temperature, and removal of impurity elements and inclusions are all carried out in the ladle and not in the tundish. All measures are taken to minimize the occurrence of macro inclusions during melt transfer from ladle to tundish. Tundish structure is made larger to avoid the decline in melt bath depth at the ladle change. A simple tundish structure and operation facilitate the hot tundish cycle without causing any additional reoxidation and slag entrainment in tundish.

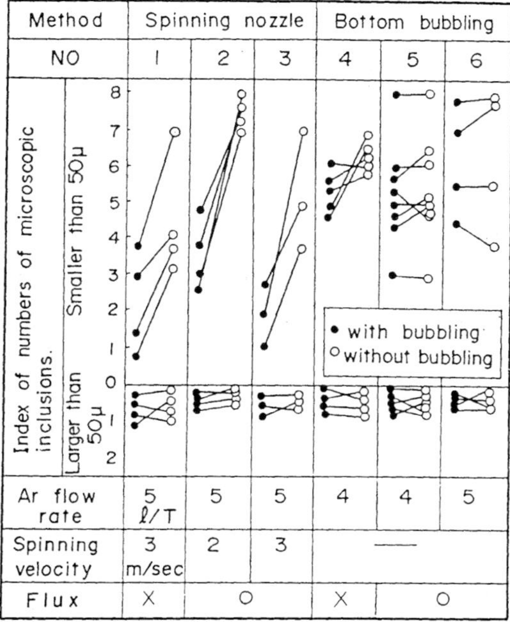

Figure 6.30: Effect of tundish gas bubbling on inclusion index. [Ref. 34]

6.7 Role of Tundish Flux

Tundish flux plays an important role in controlling the quality of the cast product. A tundish flux has the following important functions:

(1) to act as a thermal insulator by minimizing heat losses from the surface;

(2) to protect the steel melt from reoxidation by ambient atmosphere;

(3) to provide a reservoir for the absorption of inclusions;

(4) to remain inert and not be reduced and entrained by steel melt flow; and

(5) to be gentle on the tundish refractory lining and ladle shroud.

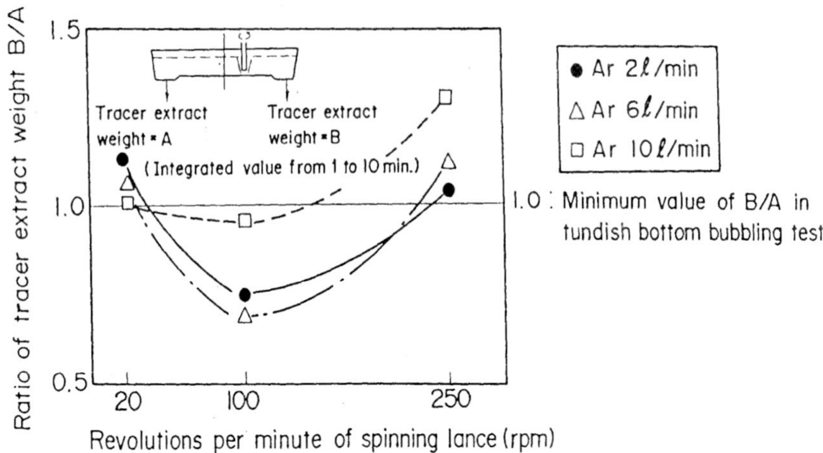

Figure 6.31: Effect of nozzle revolutions and gas flow rate on inclusion removal efficiency. [Ref. 36]

The flux should melt quickly in contact with the steel melt and should have a large solubility for inclusions. It should not contain oxides which may become a source of metal reoxidation. Assessment of metal reoxidation by oxidizing components in slag is not easy to quantify, as it depends significantly on the fluid dynamic condition of contact or mixing of slag and metal in the tundish. For a quasi-static slag/metal boundary, the rate of oxygen transfer per unit interfacial area from slag to Al-killed low carbon steel was determined by Fuchigami and Wakoh [37] and is shown in Fig. 6.32. The transfer rate depends on the amount of FeO in slag and [Al] in metal. The rate increased linearly with (FeO) in a slag containing 43-45 %CaO, 43-45 %Al$_2$O$_3$, and 5-10 %FeO for [Al] greater than 0.03 %. This shows that decreasing (FeO) in slag is an effective method of reducing the reoxidation by slag for a fully Al-killed steel melt.

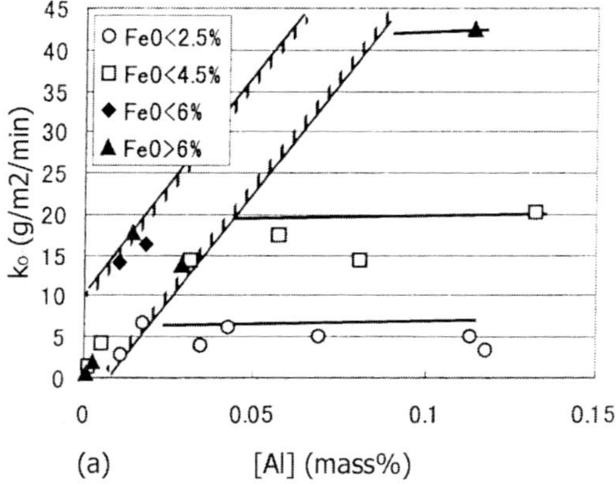

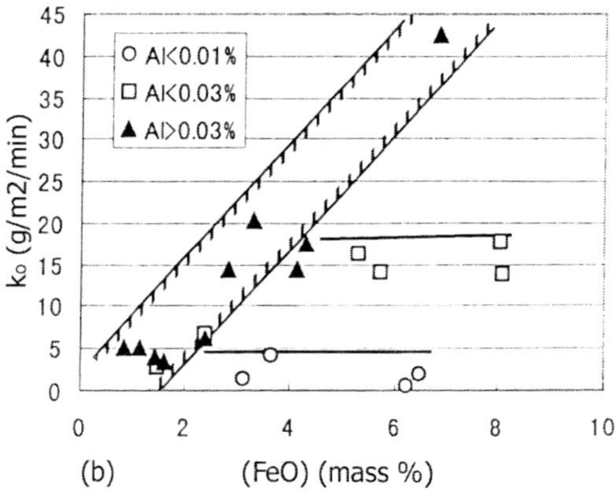

Figure 6.32: Rate of oxygen removal from slag to LCAK steel melt as a function of Al in metal and FeO in slag. [Ref. 37]

The factors which govern tundish slag performance include viscosity, melting point, interfacial tension, and inclusion solubility. The viscosity of tundish slag is an important characteristic. Highly viscous slag is slow to absorb inclusions, and highly fluid slag runs the risk of being easily entrained. Recycling of hot tundish is now becoming an important and commonly used operation, especially in Japanese steel companies (described in section 6.11). In this operation, a tundish after one sequence casting is taken to another site where the remaining metal and slag are poured out of the tundish. It is quickly repaired of any refractory damage and is returned in the hot condition for the next sequence. For this, the tundish slag should be fluid for easy and quick pouring. During one sequence casting consisting of several ladles, the volume and composition of slag in the tundish continuously change. As a consequence, the composition change results in a change in the melting point and viscosity of the slag. The melting point of the slag should be such that the slag in contact with steel should stay in a molten state for inclusion absorption, and the flux on top should stay solid for thermal insulation. Slag in liquid state has a higher rate of inclusion assimilation, whereas solid flux having gas in its pores acts as a better heat insulator. Thus, a good thickness of liquid layer and a sufficient solid layer of flux are preferred. With the metal being at approximately 1550 °C, a flux with a melting range of 1300 to 1400 °C can provide a good thickness of liquid as well as solid layers. A high interfacial tension between the slag and molten metal is preferred, as a low interfacial tension may result in entrainment of slag. Finally, a high solubility of inclusion in the slag is also a necessary requirement of the flux.

Fluxes of many different compositions have been suggested and used in the tundish. Probably, the most commonly used flux is rice hull, which contains up to 90 %SiO_2, and provides an excellent protection from heat loss. However, rice hull is very poor for absorption of inclusions, and its high silica content provides a source of reoxidation. Silica, after fluxing with reoxidation products, such as iron and manganese oxides, forms a very high viscosity liquid slag which has poor inclusion absorption ability. Fly ash has also been commonly used as tundish flux. It provides a good thermal insulation and protection from air, but contains a high fraction of SiO_2 which also becomes a source of reoxidation. Magnesia rich fluxes also are very good thermal insulators, but lack in their ability to perform other functions. Generally, tundish fluxes contain two or more of the following oxides.

$$CaO - SiO_2 - MgO - CaF_2 - Al_2O_3$$

6.7.1 *Basicity of slag*

As stated earlier, SiO_2 is a reducible oxide and may cause reoxidation of aluminum by the following reaction:

$$3\,SiO_2 + 4[Al] = 2\,Al_2O_3 + 3[Si] \qquad (6.1)$$

$$\Delta G° = -720{,}680 + 133\,T \qquad (6.2)$$

Figure 6.33 [38] shows the equilibrium curve for Eq. (6.1) at [Al] = 0.03% in molten steel at 1823 K, and the activity of SiO_2 at different basicities of CaO - SiO_2 - Al_2O_3 slag. The figure shows that a high CaO/SiO_2 ratio is needed to avoid reduction of silica from slag.

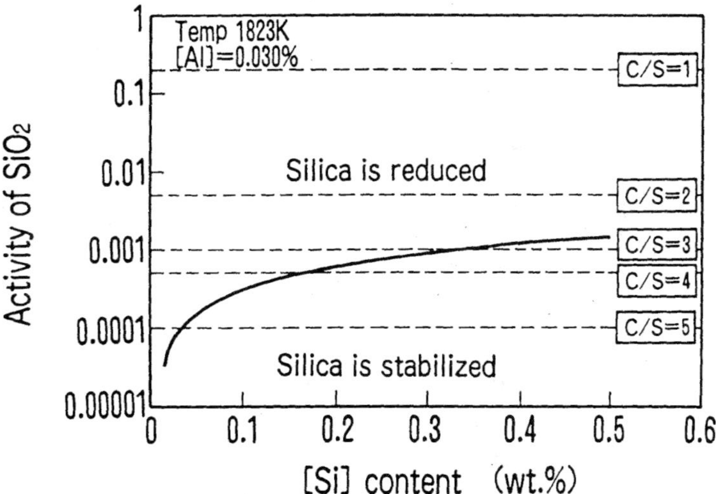

Figure 6.33: Activity of silica as a function of silicon content (C/S = %CaO/%SiO$_2$). [Ref. 38]

Bessho *et al.* [39] examined four tundish fluxes of varying CaO/SiO_2 ratios in casting ultra low carbon Al-killed steel (C < 0.003%, Al = 0.04 to 0.05%, Si <0.01%) in a two-strand caster with a 250-ton ladle and a

50-ton tundish. Composition and softening temperatures of the four fluxes are given in Table 6.2.

Table 6.2: Chemical compositions of tundish fluxes [Ref. 39]

	A	B	C	D
SiO_2 %	47.4	7.6	5.7	2.6
Al_2O_3 %	2.7	22.2	20.7	20.5
CaO %	39.5	46.2	62.5	58.2
CaO/SiO_2	0.83	6.1	11.0	22.2
Softening Temperature °C	1280	1380	1300	1410

Figure 6.34 demonstrates the change in total oxygen content in ladle after the RH treatment to the transfer to tundish as a function of the CaO/SiO_2 ratio of the flux. Although there is some scatter in the data, it is obvious that a high CaO/SiO_2 ratio is needed to decrease the total oxygen content of steel. The total oxygen content at the tundish outlet during a sequence casting is shown in Fig. 6.35, which shows that the higher basicity flux C produces steel with a lower total oxygen content in comparison to the low basicity flux A. Figure 6.36 shows the distribution of alumina clusters with a diameter larger than 0.5 mm across the thickness of slabs. Flux C resulted in an extremely low number of inclusion clusters in comparison to the slabs cast using flux A. The presence of large clusters of inclusions and higher total oxygen is due to the large amount of SiO_2 in flux A, which becomes a source of reoxidation. In Fig. 6.37, Bessho *et al.* [39] plot the activities of SiO_2 with the use of fluxes A, C, and D. When flux A is used, the activity of SiO_2 varies from 0.20 at the initial stages of casting to about 0.10 at the end of casting. In contrast, with high basicity fluxes C and D, the activity of SiO_2 was about 0.008 for flux C and 0.004 for flux D near the end of casting, which is 1/10th or 1/20th of flux A. So a flux with a basicity of 10 or higher is preferred for cleaner Al-killed steel casting.

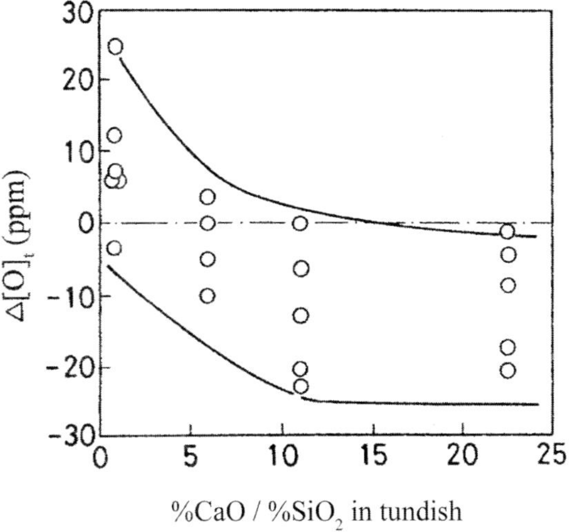

Figure 6.34: Change in total oxygen content of steel from ladle to tundish as a function of the %CaO/%SiO$_2$ ratio in the tundish flux. [Ref. 39]

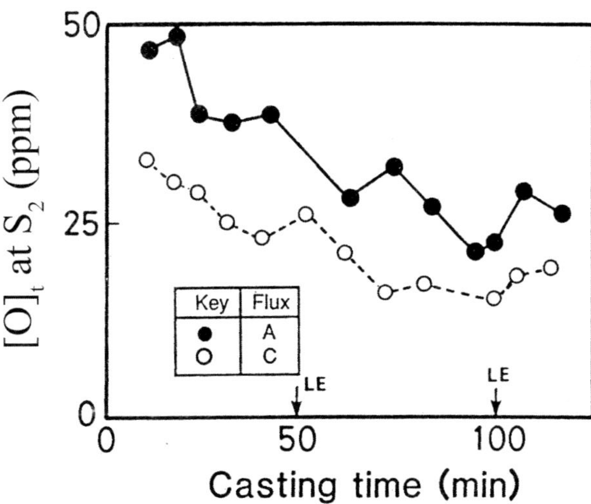

Figure 6.35: Total oxygen content as a function of casting time; (LE is Ladle Exchange or Ladle Change, S$_2$ is the tundish outlet location). [Ref. 39]

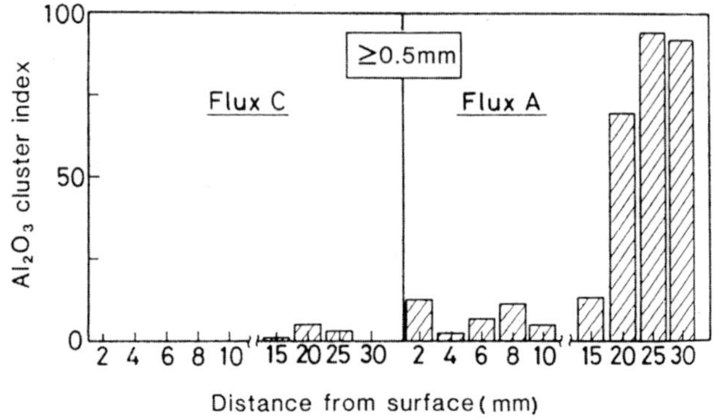

Figure 6.36: Distribution of alumina clusters across the thickness of cast slabs. [Ref. 39]

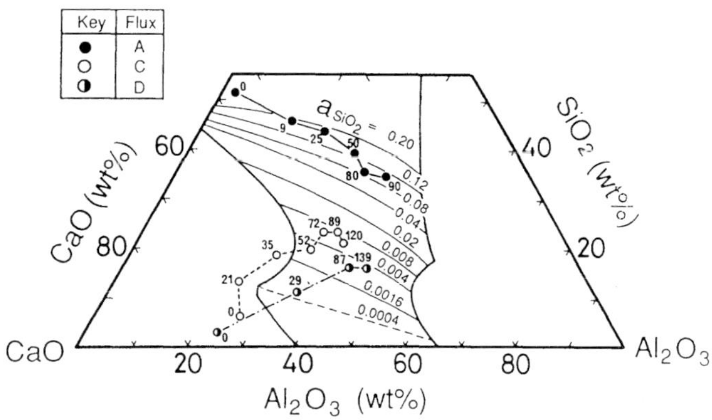

Figure 6.37: Activities of SiO$_2$ with the use of different tundish fluxes. [Ref. 39]

6.7.2 *Alumina dissolution rate*

Fukuzaki *et al.* [38] studied the dissolution of alumina in ternary fluxes. They found that the CaO - CaF$_2$ - Al$_2$O$_3$ system exhibited a better Al$_2$O$_3$ dissolution rate than the CaO - CaF$_2$ - SiO$_2$ system, and that the dissolution rate depended upon the Al_2O_3 content of the flux. Low Al_2O_3 content slags had a higher dissolution rate as shown in Fig. 6.38.

The absorption rate of Al_2O_3 in CaO - SiO_2 - CaF_2 slag containing 20% CaF_2 as a function of CaO/SiO_2 ratio is shown in Fig. 6.39 [40]. It shows that the absorption rate increases with the basicity of slag. Figure 6.40 [40] shows the influence of CaF_2 content on the absorption rate of Al_2O_3 in CaO - SiO_2 - CaF_2 slag. The rate increases with the CaF_2 content in the slag.

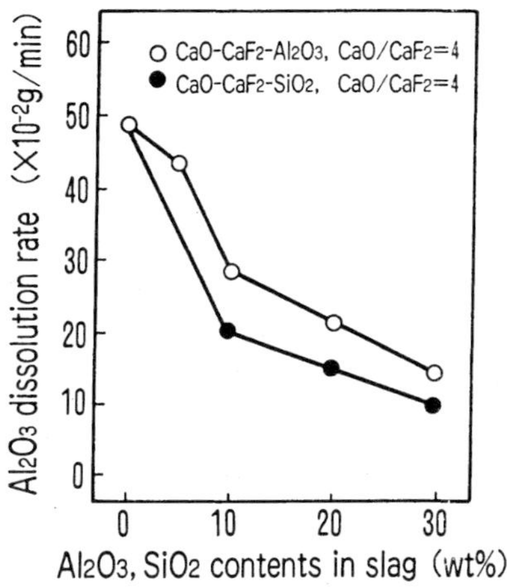

Figure 6.38: Alumina dissolution rate as a function of tundish slag composition. [Ref. 38]

6.7.3 *Slag viscosity*

Slag viscosity increases with decreasing temperature. It also decreases with increasing FeO and MnO contents in the slag. Their presence in the tundish slag is deleterious and should be avoided, as they become a source of metal reoxidation. The viscosity of CaO - Al_2O_3 - SiO_2 slag increases as Al_2O_3 or SiO_2 content is increased. Calcium fluoride is a known fluidizer, and its addition decreases the viscosity of the slag. However, the presence of CaF_2 is corrosive for basic refractory boards, and can be a source of environmental pollution after the flux is solidified and the fluorine is leached out of the solidified flux by water.

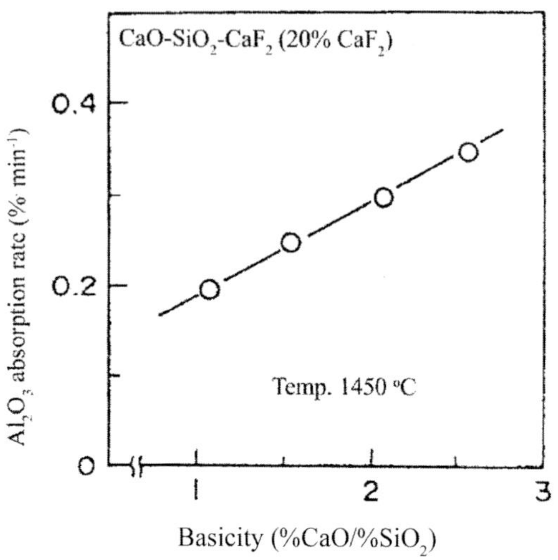

Figure 6.39: Alumina absorption rate as a function of tundish slag basicity. [Ref. 40]

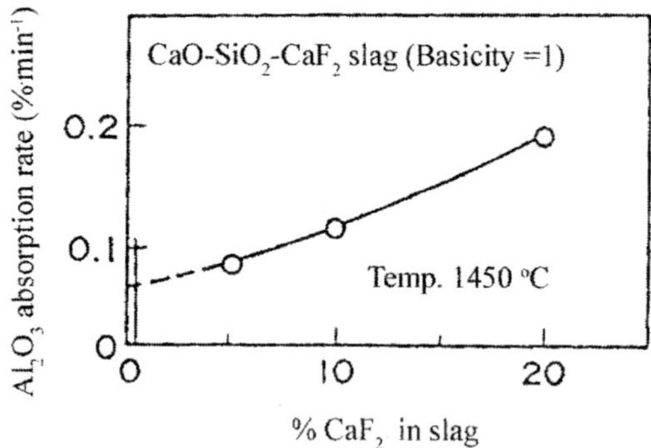

Figure 6.40: Alumina absorption rate as a function of CaF$_2$ in tundish slag. [Ref. 40]

6.7.4 *Slag composition*

Based on above discussion, it is clear that the tundish slag composition should largely depend upon the tundish operation and the grade of steel being cast. The following compositions may be representative of the range of fluxes commonly used for cleaner Al-killed steel practice:

CaO/SiO_2	> 10
Al_2O_3	5 to 20 %
CaF_2	5 to 20 %
MgO	5 to 10 %
SiO_2	5 to 10 %

6.7.5 *Metallic Al in tundish fluxes*

In sequence casting with several ladles in one tundish, slag volume in the tundish continuously increases after each ladle due to the unavoidable addition of ladle sand at the opening of a new ladle. Slag volume also increases from any carryover ladle slag, and due to the transfer of inclusions from the steel to the tundish slag. Thus, the amounts of reducible oxides, such as FeO, MnO, and SiO_2, also increase in tundish slag. These oxides, as discussed in the earlier sections, become a source of metal reoxidation. Thus, as the quantity and activity of these oxides in slag increase, the total oxygen content of metal also increases. To solve this problem, Nakatao *et al.* [41] investigated the effect of conditioning the existing tundish slag (Flux A) with a $CaO - SiO_2 - CaF_2$ flux (Flux B) containing 38 % metallic aluminum at the Kakogawa Works of Kobe Steel, which employs a hot recycle tundish. This Al metal reduced the total Fe content in slag to a low level, as shown in Fig. 6.41. Thus, high quality slabs were cast at a low cost.

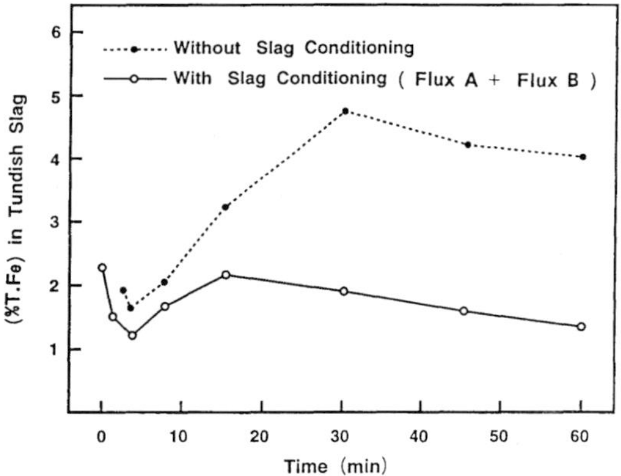

Figure 6.41: Effect of tundish slag conditioning on total iron content in slag. [Ref. 41]

6.7.6 *Tundish operation without tundish fluxes*

It should be mentioned that some quality steel producers run the tundish without any tundish flux. In this practice, the ladle to tundish melt stream is well protected from air reoxidation by Ar shrouding or a long nozzle. The tundish is covered by an air-tight lid, and Ar gas is injected above the melt surface. The ladle slag carryover is also minimized. The flux free tundish operation can prevent flux entrainment to the mold, but the operating cost is increased due to the installation of the air-tight lid and Ar injection. The tundish operation is very effective in producing clean steel if it is purged and filled with Ar gas, if ladle slag carry over is kept at a minimum, and if a clean steel supply is assured. The steel contains only a very small number of inclusions which rise and float at the free surface.

6.8 Calcium Addition

Calcium injection in the ladle and tundish has been successfully attempted by various steel companies. Some of the benefits obtainable by calcium addition to melt in the tundish are:

(1) to transform large alumina inclusion clusters into liquid calcium aluminate inclusions, which after solidification and during subsequent hot and cold rolling are elongated and fragmented into smaller sizes that do not impair steel properties;

(2) to reduce nozzle clogging and eventual blockage by alumina inclusions, sustaining a consistent melt flow rate through the nozzle and a desired melt flow pattern in the mold to maintain caster productivity and cast product quality;

(3) to convert dissolved sulfur into oxysulfide inclusions that are non-deformable during hot rolling, thus preventing the formation of deformable manganese sulfide inclusions in plate or strip; and

(4) to improve the isotropy of mechanical properties and resistance to Hydrogen Induced Cracking (HIC) that are caused by the elongated manganese sulfide during hot rolling.

Thus, calcium addition has a significant impact in improving the quality of cast steel. Van der Heiden *et al.* [42] injected CaSi wire in Hoogoven's 60-ton tundish. For calcium addition, a large volume and deeper tundish is necessary because of the high vapor pressure of calcium. They injected a 9 mm diameter cored wire containing 30 g Ca per meter, at a rate of 100 m/min. Table 6.3 compares metallurgical effects and quality improvements of a Ca-treated heat with an untreated heat.

Calcium content in the treated heat was 35 to 40 ppm, which is about the correct level. The alumina inclusion clusters were completely eliminated. The MnS inclusions of types II and III were significantly reduced, and the Hoogoven's measure of stringers in plates decreased from 18 to 4. Microprobe analysis revealed that the type III inclusions also contained CaS, CaO, and Al_2O_3 which hardly elongate during hot rolling. Excessive addition of Ca is deleterious to properties of steel because excess Ca reacts with slag and refractory to form oxide, some of which is entrained in the steel melt as CaO based inclusions.

Tolve *et al.* [31] at CSM Italy injected CaSiBa alloy filled in 9 to 11 mm wire. The wire was coiled in a spiral shape of 250 mm in diameter and 100 mm pitch. Using feed rates of 20 to 30 m/min, 0.15 to 0.23 kg of Ca/ton of steel was introduced. The resulting slabs were rolled into 16 mm thick plate. The Charpy-V tests showed that the transverse Charpy-V values were very close to the longitudinal ones, and thus, optimum isotropy was achieved in the treated steel. The inclusions in the treated plates were found to be globular-shaped, consisting mainly of calcium sulfide nucleated on calcium aluminates. The inclusions in the untreated plates consisted of manganese sulfides and clusters of alumina.

Table 6.3: Metallurgical effects of Ca-treatment of steel in a tundish. [Ref. 31]

	Ca-Treated	Reference Heat
Sulfur	0.007 %	0.007 %
Stringers in plates	4	18
Reduction in area of plate	63 % in 15 mm thick, and 73 % in 40 mm thick plate	50 % in 25 mm thick plate
Recovery of Ca	8.5 %	-
MnS-type III in segregation zone	$14.5/cm^2$	$25.4/cm^2$
MnS-type II in segregation zone	$0.9/cm^2$	$39.5/cm^2$
Ca-level	35~40 ppm	-
Al_2O_3 clusters in slab	0	$0.8/cm^2$

Kitamura *et al.* [43] at Kobe Steel made a calcium alloy addition to steel and found that the cast structure had an increased proportion of equiaxed grains, and the morphology and composition of sulfide inclusions were modified. Yoshii *et al.* [44] at Kawasaki Steel found that the Ca-treated steel had a significantly reduced number of manganese sulfide inclusions as well as large oxide inclusions with the addition of 50 ppm Ca (shown in Fig. 6.42). They also found that the number of inclusions in the transition slabs cast during ladle change was reduced, and the overall cleanliness was improved in the treated steel.

Figure 6.43 shows a simple method of CaSi addition to the ladle to tundish stream, which was described by Haida *et al.* [45]. The method was found to be effective in making alloy additions, and was successfully practiced to produce HIC resistant line pipe steel. The yield of Ca is about 15 to 20%, which is reasonably high despite the simple method of addition, since no oxidizing slag existed in the inlet compartment of the tundish where the CaSi addition was made. They found that the sulfide shape control was insufficient for the atomic concentration ratio (ACR) of effective Ca to S of 0.2, and became satisfactory at ACR of 0.4 except for center-line segregation.

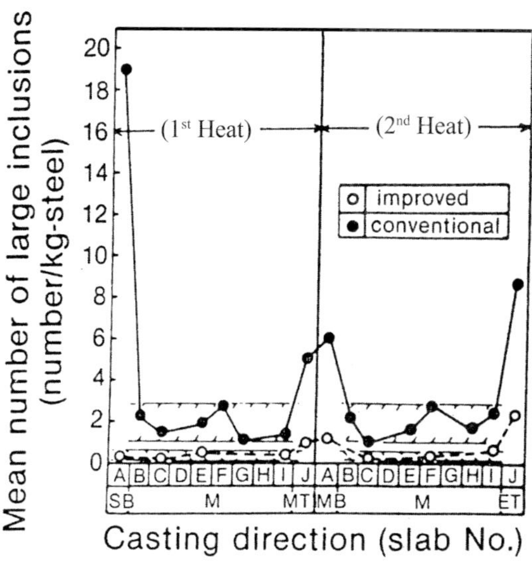

Figure 6.42: Effect of calcium treatment on large inclusions. [Ref. 44]

It must be noted, however, that alumina clusters and dissolved sulfur in Al-killed steel melt can be easily converted into liquid lime aluminates and solid lime oxysulfides by Ca-addition in the ladle. This is because stirring steel melt in the ladle is much better than in the tundish. Better stirring assures better dispersion of Ca, more homogenous conversion of alumina into lime aluminates and dissolved S into calcium sulfide/oxysulfide, and improved removal of lime aluminates and calcium sulfide/oxysulfide to the top slag. In order to achieve better yield of Ca in the ladle, oxidizing components in the ladle slag must be minimized, and less reducible refractory must be used for the ladle lining. A Ca-wire addition to the tundish causes significant turbulence at the tundish slag/steel melt boundary, which may entrain tundish slag into the steel melt as macro inclusions. Some fraction of the entrained inclusions is inevitably carried over into mold and trapped in cast strand.

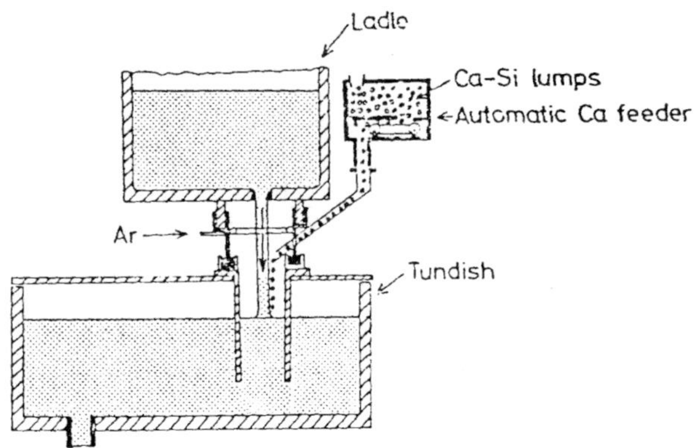

Figure 6.43: A simple method of CaSi addition. [Ref. 45]

Thus, the addition of Ca to a tundish should be regarded as the last measure, when nozzle clogging cannot be avoided otherwise, *e.g.* under the following circumstances:

(1) when a ladle refining facility is not available to clean the melt, and hence high alumina inclusion content melt has to be transferred to the tundish. In such a case, ladle slag is usually high in FeO and MnO, and hence Ca added to the steel melt is consumed in reducing these oxides, making the Ca yield too low and uneconomical;

(2) when reoxidation of the melt during the melt transfer from ladle to tundish and in the tundish is not satisfactorily prevented due to operational constraints, and hence, additional alumina clusters are generated in once cleaned melt; and

(3) when the nozzle aperture is inevitably small and easily clogged, such as in billet casting and thin slab casting.

In fact, high quality Al-killed clean steels are now cast in blooms and slabs without Ca-addition by a combination of the following measures:

(1) use of a ladle refining process;

(2) strict prevention of reoxidation and ladle slag carryover during melt transfer from ladle to tundish;

(3) a large tundish covered by a basic tundish slag or Ar atmosphere; and

(4) an Ar-injected large bore submerged entry nozzle.

Morphology control of alumina clusters and sulfides is done after $[O]_t$ and S are minimized by ladle refining. Then Ca is added to the ladle with basic top slag low in FeO and MnO content. Ca-addition to the tundish is still mostly practiced for billet casting and thin slab casting of Al-killed steel.

6.9 Sequential Casting of Different Grades

Two approaches are common in the steel industry for sequentially casting different grades of steel. In casting radically different grades, normally a tundish used for the old grade is removed and a new tundish is quickly (say in less than 5 minutes) brought in and the casting continues in the same mold. In this case, some mixing of the two grades takes place in the mold. Grade separators and strand links are commonly used in the mold to minimize the length of the transition grade of steel cast. The second approach is generally employed when the chemistries of the two grades are compatible and not too dissimilar. In this approach, the melt level in the tundish is lowered to an acceptable level (which does not cause slag entrainment due to vortexing) and the new grade is poured into the tundish. Thus, the same tundish is used for casting two different grades. The casting in the mold continues without any interruption, but results in some transition length.

6.9.1 *Use of grade separators and strand links*

Various designs of grade separators and strand links have been developed and used by different steel companies. The addition of a grade separator causes local freezing of the old grade metal before the new grade is poured into the mold. This prevents the mixing of the two grades and minimizes the transition length. Figure 6.44 schematically shows a design developed and used by Bethlehem Steel Corporation [46]. Figure 6.45 schematically shows the sequence of events taking place during the casting of two different grades. It is clear from the chemistry of steel near the middle of the strand on both sides of separator, plotted in Fig. 6.46, that the new grade has not penetrated below the separator, thus confirming the effectiveness of a grade separator. This total grade change operation took less than a minute. A total length of 1.2 m was sufficient to remove any defective material, which consisted of a mixed composition zone, the separator, and the shrinkage cavity below the separator.

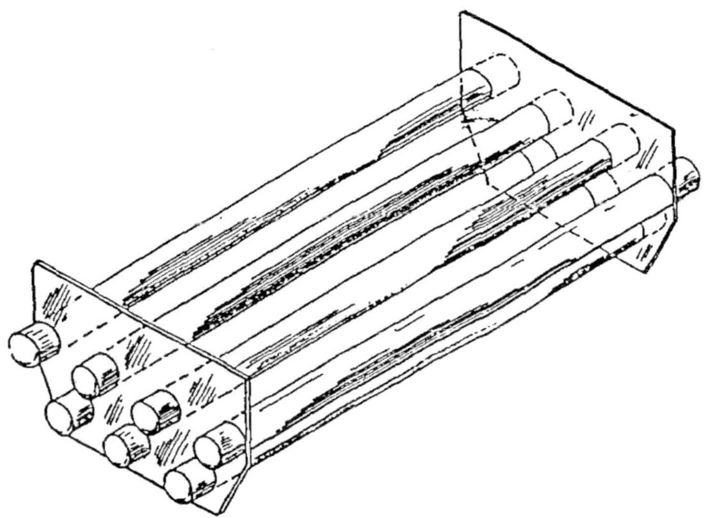

Figure 6.44: Schematic of a grade separator. [Ref. 46]

Strand links are placed at the surface of the steel, where half of the link is frozen in the old grade by keeping for a sufficient time. The upper half of the link is solidified in the new grade of steel. The primary purpose of a strand link is to form a sound mechanical link between the two grades of steel. Any 'dog-bone' or 'ring' shaped device is suitable for this purpose. Figure 6.47 schematically shows the application of a strand link for the casting of two grades, which was also developed at the Bethlehem Steel Corporation [46]. The discard length was about 1.2 m, the same as in the separator case. The casting interruption is about 5 minutes with the use of the strand link. During this period of casting interruption, the metal in the mold is caused to move down very slowly to avoid sticking to the mold. However, such an operation may cause excessive cooling of the slabs that imposes unfavorable pressure on the bearings and roll of the caster. Thus, grade change is the last measure after all efforts have been taken for scheduling the casting sequence to avoid the grade change.

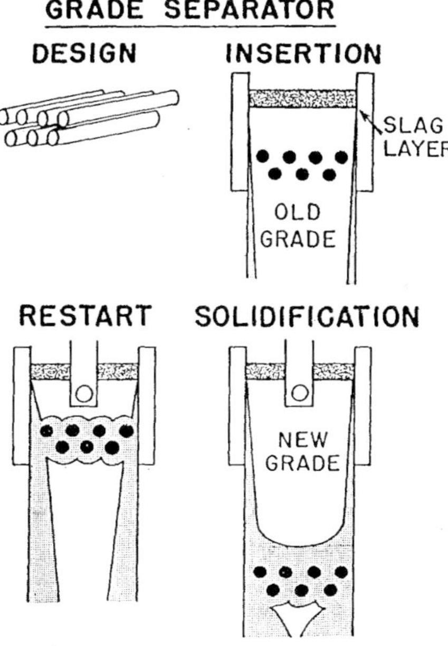

Figure 6.45: A grade separator during grade change. [Ref. 46]

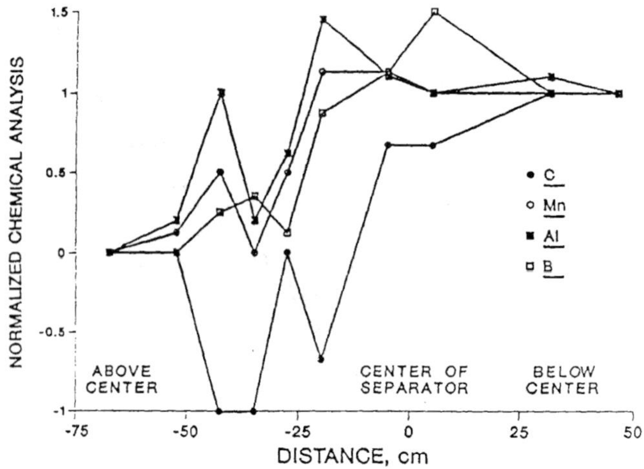

Figure 6.46: Metal composition change during grade change. [Ref. 46]

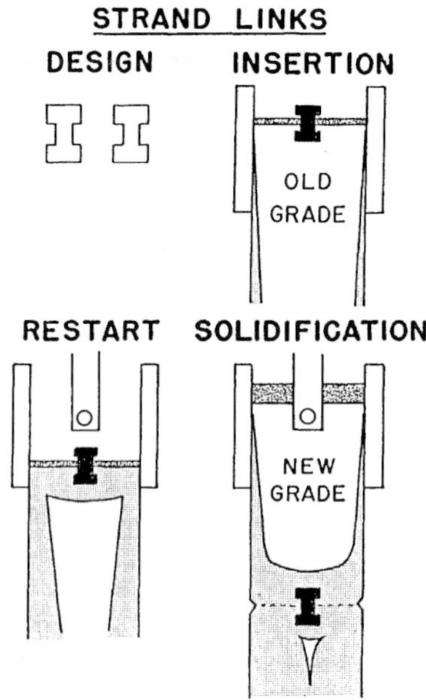

Figure 6.47: Use of strand link during grade change. [Ref. 46]

6.9.2 *Casting of different grades using the same tundish*

As stated above, the melt in the tundish with the old grade of steel is lowered to a predetermined level before pouring the new grade. The new grade is usually poured at a rate which is much greater than the steady state casting rate. This is done to fill the tundish to its normal operational level as quickly as possible. During this operation, the casting rate may either be reduced or kept constant. Thus, the mixing in the tundish causes some transition length of steel whose chemistry does not fall within the specification of either steel grade. Several researchers [*e.g.* refs. 47-49] have employed water and mathematical modeling studies, and plant trials to understand the influence of various operational parameters on the transition length.

Diener *et al.* [47] published a detailed study where they used a *tanks-in-series* model to describe mixing in a slab caster of Hoesch Stahl AG in Germany. Their study developed a grade change strategy whereby they drained a 28-ton tundish to 6 tons prior to a rapid refill to its steady state operational level. This practice reduced the average transition slab length from 9 m to 5 m.

Burns *et al.* [48] studied mixing in ARMCO Steel's Ashland slab caster using a full scale water model. The tundish was drained to different depths before refilling. They demonstrated an important fact that a transition from a narrow to wide carbon specification resulted in a shorter transition length. The effect is better illustrated in Fig. 6.48, which displays two plots of chemistry versus slab lengths. The exponentially increasing line is for a transition proceeding from a narrow to wide specification (0.03-0.05% C to 0.13-0.17% C). The terms "narrow" and "wide" define the absolute range of the specification, with 200 ppm in the former case and 400 ppm being the latter. The exponential decay line is the reverse, and represents a transition from a wide to narrow chemistry range. In comparing the right-hand tails of the exponential transition in Fig. 6.48, it is seen that the transition proceeding from narrow to wide specification enters into the specification earlier, resulting in a reduction of 3.5 m transition length for the example presented here.

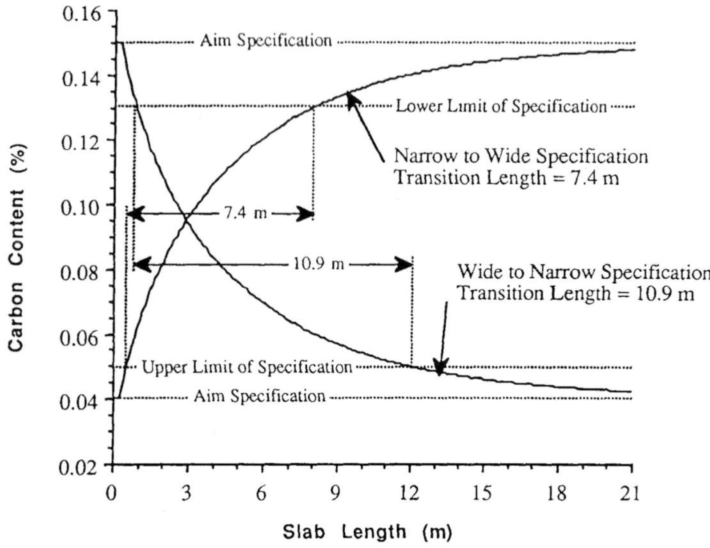

Figure 6.48: Metal composition change during grade change. [Ref. 48]

Based on their results with the water model [48], a mathematical model was developed which was capable of relating the rate of chemistry changes with time of casting of new grade. The model was verified using multiple mold and slab samples during the grade transition at the caster. Using this model, the transition grade was limited to only one slab, where previously two slabs were discarded.

Damle and Sahai [49] performed a mathematical modeling study, where they found that in addition to the depth of the emptying tundish, the rate of filling back to the normal steady state level also influenced the transition grade. Thus, filling very rapidly may not necessarily result in the least mixed grade. This optimum rate of tundish filling depended upon the chemistry specification range and the design of the tundish. Thus, the transition grade can be further reduced by employing a combination of filling rates.

6.10 Tundish Refractory

Refractories in a tundish should provide adequate thermal insulation, and the layer of refractory in contact with the molten steel and slag should not interact adversely. Physical and chemical interactions with slag may be more severe and thus, require very high quality refractory. Some areas of a tundish also require higher thickness or a special refractory. These include the impact pad, where the molten steel stream impinges on the refractory, and near the plasma torch, where high temperatures are generated. Any refractory wear is detrimental to the quality of steel, as it causes formation of exogenous inclusions.

The steel shell of a tundish has a permanent lining, which generally consists of high silica. This lining provides very good thermal insulation. The next thickest layer is the working lining, which may be brick or castable lining, and this refractory may be zirconia, high alumina, or magnesia based. Large tundishes of high productivity utilize alumina-silica castable monolithics for durability, labor, and cost reasons. Finally, in most cases, this refractory is coated with high magnesia castables often by high speed gunning. These monolithic castables should be dried, sintered, and preheated before being put in service to prevent explosion spalling. In special cases, magnesia board is used instead of high magnesia castables. Taguchi *et al.* [50] have described the use of duplex boards consisting of a high magnesia front board and a high silica back board. The composition and physical properties of the board are given in Table 6.4. They compared the heat loss and relative total cost of this board with castable lining, and found that the duplex board provides better thermal insulation and that total cost is about 73% of the castable refractory. In either case, the refractory material of the working face in contact with the steel melt should be high in MgO, with a minimum amount of reducible binder to prevent oxidation of Al in the melt. The choice of coating or board depends on local requirements of labor cost, refractory cost, life of the refractory, etc. Gunning of magnesia castables is now common for high productivity casters in major integrated steel mills [51].

Table 6.4: Composition and properties of duplex board. [Ref. 50]

	% MgO	% SiO$_2$	Ignition Loss, %	Density kg.m^{-3}	Thermal Conductivity kcal.m^{-1}hr^{-1}K^{-1}
Front Board	80	<4	13	1700	0.7
Back board	<1	82	7	1000	0.2

The tundish is generally preheated to about 1200 °C before pouring the first heat, but in limited cases where preheating is difficult, it can also be used in the cold state at the expense of temperature loss and quality. Even in the cold state, the nozzle and stopper rod are preheated to avoid any freezing of metal. Metal temperature of the first heat in cold tundish practice drops significantly, and thus, requires a higher superheat. Japanese companies generally use a preheated tundish practice to avoid any freezing and to keep higher steel temperature. Some companies in North America, however, used a cold tundish practice.

Figure 6.49 shows a schematic of Kobe Steel's 80-ton tundish [51] and its lining profile. The tundish is sealed with a cover, and is plasma heated. It can be seen that the refractory in the area (marked No. 2 in the figure) closest to the plasma torch location has 80% alumina refractory. The impact region (marked No. 9) has a thicker refractory and has a high alumina precast block. The tundish is recycled in a hot condition without any refractory maintenance up to 150 times. With quick refractory maintenance, it is used for 3 or 4 such cycles of 150 heats. Thus, a total of 500 heats are cast without complete relining of the tundish.

6.11 Recycling of a Hot Tundish

After a sequence casting, the removal of skull from a tundish sometimes damages the working lining. This tundish lining repair is very expensive in terms of refractory cost and is a labor intensive task. So, many steel companies, especially in Japan, have adopted hot tundish recycling, pioneered by Kobe Steel [52]. In this operation, soon after one sequence casting, the tundish is tilted to pour out any remaining metal and slag in it, followed by an automatic change of the old pouring nozzle with a new and preheated nozzle, replacing the slide gate valve, and repairing any refractory damage. Finally, the tundish may be preheated to a desired temperature before returning for another casting sequence. In a hot tundish recycling operation, it is not necessary to perform all the above steps. The preheating step is generally avoided as it causes oxidation of any remaining metal in the tundish, which contributes significantly to the formation of non-metallic inclusions in the next sequence. Refractory repairs are only performed if they are deemed necessary. As Fig. 6.50 shows, therefore, the tundish may return to operation after pouring the remaining slag off, or it may return after the slide gate valve change, or it may go for refractory repair and preheating before returning for the next casting sequence. The hot recycle process has reduced 8 hrs of tundish preparation time for conventional process to only 25 min, and reduced refractory cost to less than one tenth.

Shirai [53] and Isono *et al.* [54] described the hot tundish recycling operation at the Nippon Steel Corporation's (NSC) Hirohata Works. The use of a hot tundish has improved the quality of steel cast during the unsteady state casting period, and has cut down the refractory and personnel costs significantly. Thus, the operation has contributed to the production of high quality steel at lower cost. Figure 6.51 shows the flow diagram of the hot tundish recycling operation. Within 2 minutes of the end of a cast, residual steel and slag are discharged in a slag ladle by opening a flapper device at the bottom of the tundish. After pouring the slag off, the used immersion nozzle is pushed out to an immersion nozzle change car by the stopper rod used in that cast. The nozzle change car cleans the tundish well and inserts a new preheated nozzle, which completes the tundish maintenance. The tundish is then returned for operation in the next sequence. Thus, the important features of hot tundish recycling are as follows:

(1) Steel remaining in the tundish is discharged quickly to avoid any build up of steel and slag in the tundish;

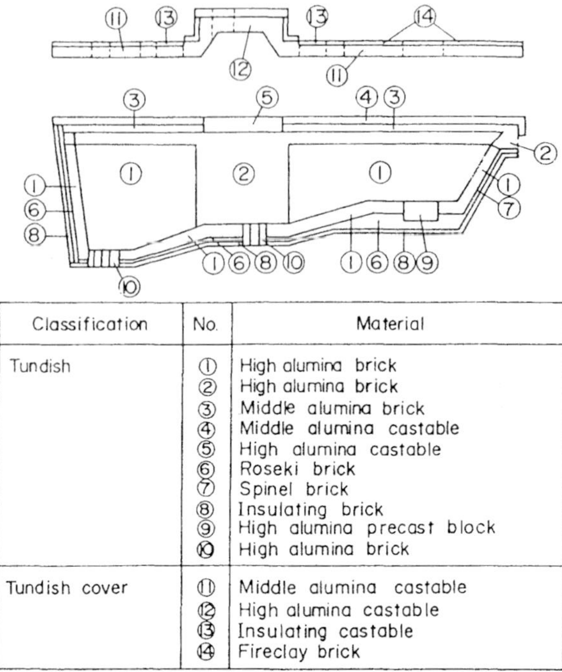

Classification	No.	Material
Tundish	①	High alumina brick
	②	High alumina brick
	③	Middle alumina brick
	④	Middle alumina castable
	⑤	High alumina castable
	⑥	Roseki brick
	⑦	Spinel brick
	⑧	Insulating brick
	⑨	High alumina precast block
	⑩	High alumina brick
Tundish cover	⑪	Middle alumina castable
	⑫	High alumina castable
	⑬	Insulating castable
	⑭	Fireclay brick

Figure 6.49: Schematic of Kobe Steel's 80-ton tundish and its lining profile. [Ref. 51]

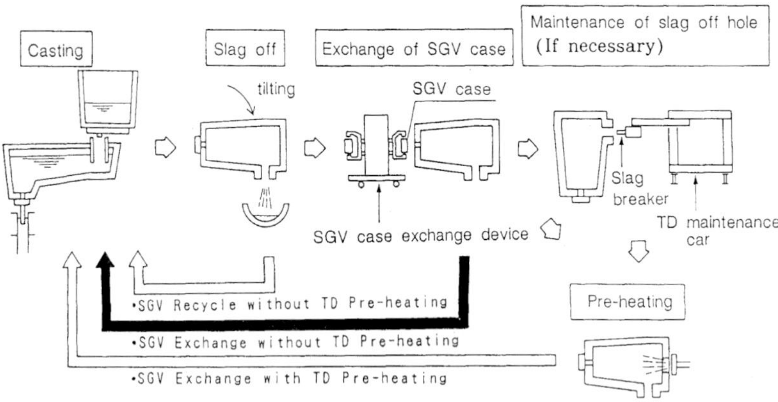

Figure 6.50: Recycling operation of a hot tundish (SGV: Slide Gate Valve, TD: Tundish). [Ref. 52]

(2) Use of a one-piece immersion nozzle prevents any air entrainment and protects the steel from reoxidation; and

(3) The tundish is operated without any preheating. The on-line maintenance is completed in less than 40 minutes for the preparation of the next sequence. Thus, preheating is not needed, and contamination of the steel is avoided.

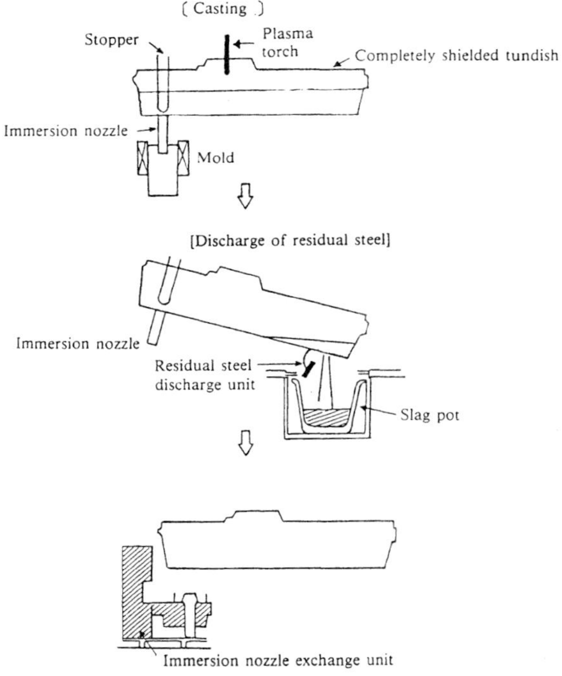

Figure 6.51: Flow diagram of the hot tundish recycling operation. [Ref. 53]

Following are some of the operational results of the hot recycling tundish.

6.11.1 *Slag crystallization and build-up*

During the continuous use of a tundish, the high melting point materials crystallize out of the tundish slag when the tundish melt temperature is low towards the end of a cast. This crystallization lowers the fluidity of slag, causing a poor discharge. Thus, some slag is left in the tundish.

This problem is corrected by continuing the plasma heating of the tundish to the end of the cast to prevent any metal temperature drop. Secondly, a tundish flux is added to lower the crystallization temperature towards the end of a cast and to facilitate the discharge of slag from the tundish.

6.11.2 *Temperature drop of tundish*

Figure 6.52 shows the tundish brick surface temperature with time after the end of a cast. The brick surface temperature stays hotter than the conventional tundish preheating temperature even after 6 hours in a completely shielded hot tundish. Since the recycling time of the hot tundish is of the order of 40 minutes, the tundish temperature may be as high as 1450 °C. This significantly improves the quality of the cast product in the early stages of casting. The quality of slabs cast during the

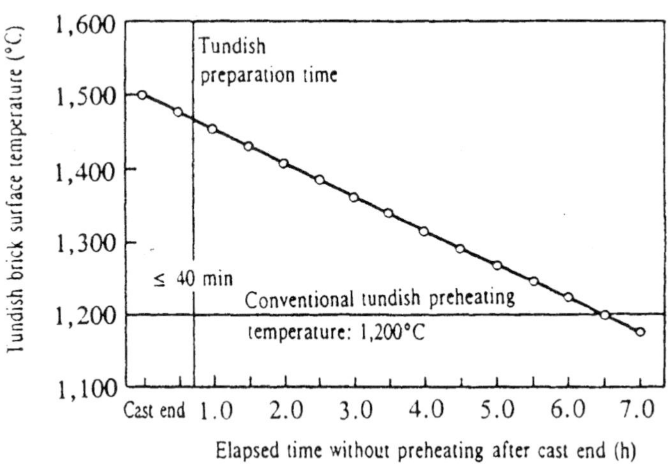

Figure 6.52: Tundish brick surface temperature change with time. [Ref. 53]

unsteady state period was found to be comparable to the steady state slabs cast using a conventional tundish.

6.11.3 *Number of heats and refractory cost*

A continuous operation of 561 heats was obtained using one tundish lining at the Nippon Steel Corporation. The refractory cost was reduced to less than 30 % of the conventional tundish refractory cost per ton of steel cast.

6.12 Starting and Ending a Sequence

6.12.1 *Starting a sequence*

At the start of a sequence, a preheated tundish is brought on the tundish car to the casting station and set in place on the molds. The ladle is attached to a long nozzle or to an Ar shrouding pipe without a long nozzle, for protecting the metal stream during its transfer from the ladle to the tundish (see Fig. 6.2). The ladle is lowered onto the tundish, and then the nozzle is opened to transfer the metal to the tundish. The ladle well is filled with a filler sand to prevent freezing of the steel, to protect the sliding plate, and most importantly to secure a free opening, when the slide gate is opened. A silica-based synthetic sand, with some low melting temperature materials to prevent thickening of the solid sintered layer of the sand, is utilized as the filler material. The composition, grain size distribution, water content, thermal expansion, etc., are tailored to match the operational conditions, in particular, the holding time of the melt in ladle. Upon opening of the ladle, the sand flows into the tundish along with the metal. The addition of silica-based sand to the metal becomes a source of reoxidation. Preventing the fall of sand into the tundish by blowing it off was reported to decrease the occurrence of macro inclusions, although this practice is possible only for Ar shrouding pipe. Tanaka *et al.* [55] at Nippon Steel Corporation tested the performance of various non-silica sands. They found that a double-layered sand filler performed better than SiO_2 sand alone. This is shown in Fig. 6.53 with SiO_2 at the top part, which is sintered in contact with steel, and Al_2O_3 or Al_2O_3 - MgO spinel in the bottom part, which does not sinter and helps in free opening of the ladle. Use of this double-layered filler sand is found to improve the number of free openings of the ladle and to improve the quality of cast steel.

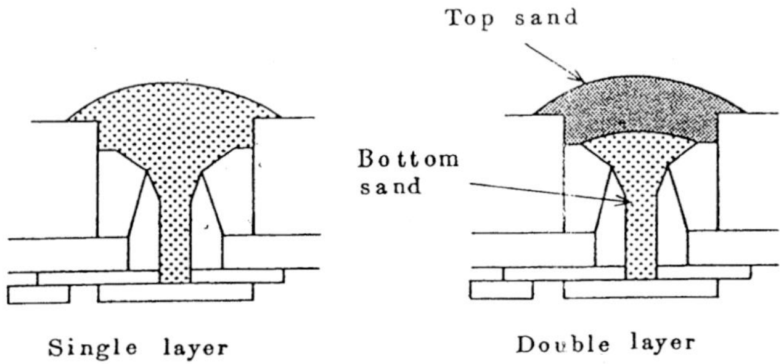

Figure 6.53: Single- and double-layered filler sand arrangement. [Ref. 55]

The slide gate is then opened to let the ladle melt flow into the tundish. During filling, a tundish flux is gently added to cover the melt surface. The flux addition is done manually or automatically by a machine. The melt continues to fill the tundish with its surface completely covered by the flux.

In tundish flux-free operation, the tundish is covered with a gas-tight lid, and the lid/tundish space is flushed with Ar gas. The ladle is set on the Ar shrouding pipe, and ladle melt is allowed to teem into the tundish. A long nozzle is not used as the ladle shroud but an Ar gas shrouding pipe is used in the flux free operation. The rest is the same as above, except for the tundish flux addition.

Some air reoxidation of steel and entrainment of the sand by steel melt impinging at the tundish are unavoidable at the ladle opening, despite Ar flushing in the tundish. It is, therefore, necessary to let the macro inclusions generated here float to the surface, and to prevent any transfer of the surfaced inclusions to the mold by vortexing. It is also necessary to reduce any excessive temperature drop of the flowing melt.

Accordingly, the melt is held in the tundish for some time, until a critical depth (above 500 mm) is reached, before the melt is allowed to enter into the mold. This procedure is called a filled start. Sand packing of the nozzle well in the tundish, over the closed slide gate valve, must be avoided, since the packed sand transfers into the mold and may get trapped in the solidifying metal as macro inclusions. In modern casters, therefore, filler sand is replaced by Ar gas injection into the tundish nozzle through thin channels that are made through the slide plate. Figure 6.54, taken from Kobe Steel's works [56], schematically shows the Ar gas injection, which prevents the freezing of metal at the nozzle and slide plate. Kobe Steel's experience [56] with gas injection at the tundish nozzle shows that opening the tundish nozzle at deeper depths causes fewer inclusions in steel. Fig. 6.55 shows that the number of inclusions in steel with the start of the cast after 10 minutes of filling is smaller in comparison to the start of casting after 5 minutes of filling. Since the residence time of liquid steel is greater and vortexing is avoided when it is filled to higher depth, it results in the casting of cleaner steel. They suggest that the 80-ton tundish should be filled to about 50 tons of metal before first opening the tundish.

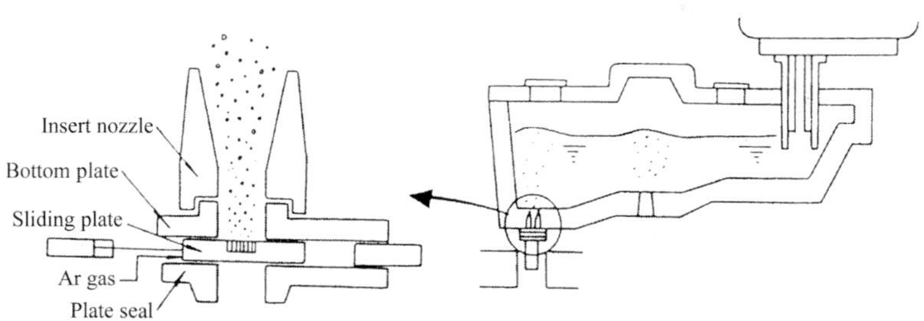

Figure 6.54: Argon gas injection in tundish nozzle. [Ref. 56]

After the end of one ladle, another ladle is quickly brought in and opened for the continuation of casting. The metal level is brought to the normal operational depth by full opening of the ladle, and then the rate is reduced to a normal steady state flow. This ladle change period is called the non-steady state period, and casting during this period generally has more inclusions than the normal or steady state casting. During this period, the withdrawal rate of the cast slab is, generally, reduced. With a deeper tundish, such as Kobe Steel's 2 m deep tundish, the casting rate during the non-steady state period is maintained the same, and as a result of the large depth, the quality is found to be much better (see Fig. 6.17). This is one of the major reasons why a large tundish is preferred for high throughput rate casting of clean steels.

In advanced casters, the whole sequence of the starting operation is fully automated by a combination of computer programming and robotics.

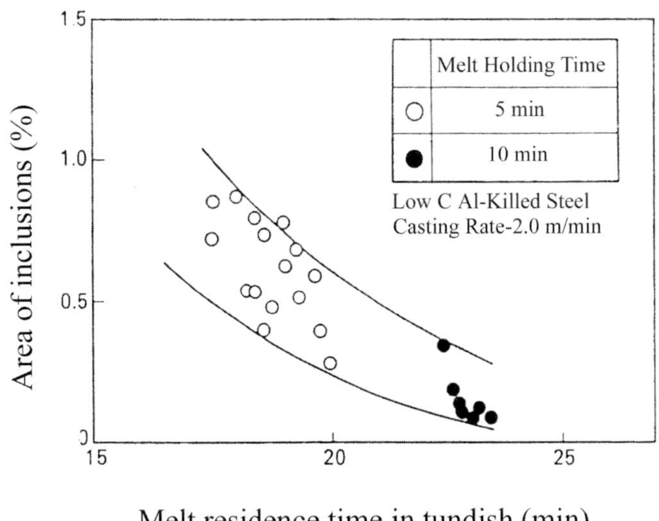

Figure 6.55: Effect of mean residence time on the area of inclusions. [Ref. 56]

6.12.2 *Ending a sequence*

It is a normal practice, towards the end of the last ladle of the given sequence, to reduce the flow from the tundish to the mold, so the casting speed in the mold is reduced. This is done to avoid any transfer of tundish slag to the mold. Depending upon the grade and quality of the steel being cast, some metal may even be left in the tundish, and casting is stopped. The slide gate is closed after completion of the metal transfer to the mold. At this time, the top surface of the metal in the mold is frozen by adding solid steel to the molten metal. The slab withdrawal speed in the mold is again increased, and the last portion of the cast slab is withdrawn from the mold.

The slower flow rate from the tundish to the mold increases the residence time of metal in the tundish, and results in large temperature drop of metal in the tundish. Usually, the metal cast in the last stages has a higher number of inclusions and cast strand shows poor subsurface quality. Kanda *et al.* [57] of Kashima Steel Works at Sumitomo Metal Industries successfully tried casting the entire metal at the normal casting flow rate and withdrawal speed, which is shown in Fig. 6.56. Thus, as shown in Fig. 6.57, the temperature drop is predicted to be much less than the conventional method. The surface of metal in the mold also was not intentionally frozen before pulling the last portion of the slab out of the mold. As shown in Fig. 6.58, this method resulted in improved quality of the cast steel in terms of inclusion content and subsurface defects, and higher caster productivity.

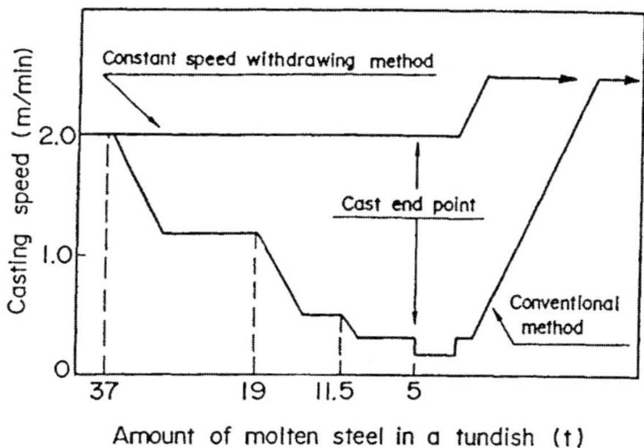

Figure 6.56: Casting at normal flow rate at Sumitomo Metal Industries for improved quality of cast steel. [Ref. 57]

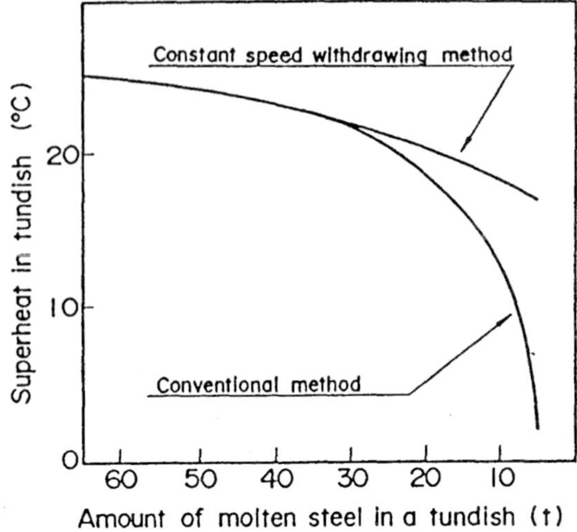

Figure 6.57: Comparison of melt temperature drop in constant speed withdrawal and conventional methods. [Ref. 57]

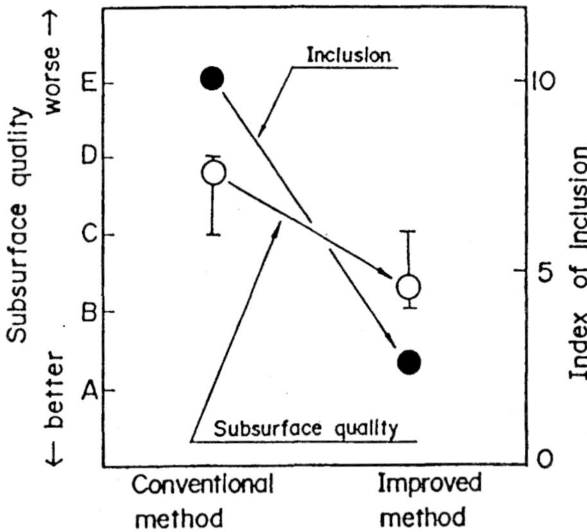

Figure 6.58: Comparison of steel quality in constant speed withdrawal and conventional methods. [Ref. 57]

6.13 Tundish to Mold Melt Delivery and Nozzle Clogging

6.13.1 *Melt delivery system from the tundish to the mold*

The tundish melt is fed through the tundish nozzle and submerged entry nozzle (SEN) to the mold. A stopper rod or slide gate placed between the tundish nozzle and the SEN controls the feed rate. The combination of a stopper rod and the tundish nozzle is often used for casters that deal with small heat size and cast a limited number of heats in a sequence for cost and durability reasons. Ar gas is injected either from the stopper rod or from a gas sleeve in the tundish nozzle, as shown schematically in Fig. 6.59, to reduce nozzle clogging caused by the accretion build-up of alumina inclusions in casting Al-killed steels. To prevent air ingress from the joint at the tundish nozzle and SEN, an integrated nozzle without the joint or single body nozzle with inset gas sleeve is often utilized. During casting heat, uniform contact between the stopper head and the upper surface of the tundish nozzle tends to be lost due to irregular wear of the head or nozzle top refractory by melt flow, and the flow control becomes erratic. To reduce the wear, suitable refractory materials with desired grain size distribution are selected. The tundish nozzle and the stopper rod head are formed and sintered from this refractory mix. Also, a single body nozzle with a durable cap is employed to achieve a longer lasting uniform contact. These are shown schematically in Fig. 6.60. Flow control by a stopper rod is generally favored by some special steel casters and European casters.

In large-size high productivity casters, a slide gate is commonly used because of its better durability and convenience for on-the-fly automatic exchange of the SEN during sequential casting without casting interruption. A two-piece slide gate consists of a top plate attached to the tundish nozzle and a slide plate to which the SEN is attached. Both the top plate and the slide plate have a through hole. A three-piece slide gate assembly has a top plate attached to the tundish nozzle, a slide plate with a bore, and a bottom plate to which the SEN is attached. The two-piece slide gate controls the feed rate by moving the slide plate with the attached SEN, while the three-piece type does the same by moving only the slide plate. Each type has its own advantages and disadvantages. For casting clean steels, however, the latter type is better because the central axis of the SEN does not move in the mold. Accordingly, symmetry of melt flow exiting from the SEN is maintained to facilitate macro inclusion flotation in the mold. However, more attention is required for the latter type to prevent air ingress from the slide plate/SEN contact.

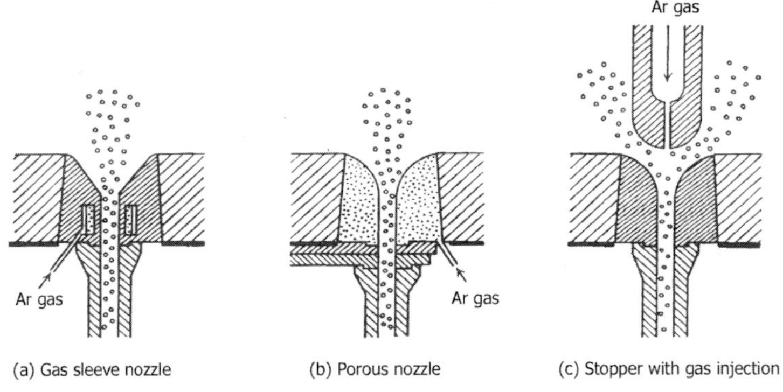

Figure 6.59: Gas injection from (a) gas sleeve, (b) porous nozzle, and (c) stopper rod. [Ref. 58]

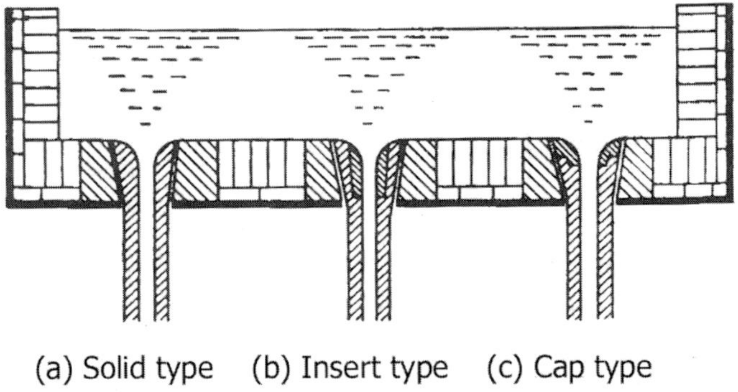

Figure 6.60: Single body nozzle with a durable top (a) solid type, (b) insert type, and (c) cap type.[Ref. 58]

The withdrawal rate of the casting and the meniscus level of steel melt in the mold during steady state casting are kept constant for quality and productivity reasons. The melt feed rate from the tundish to the mold is kept constant to meet the withdrawal rate requirement, first by maintaining a constant bath depth in the tundish and second by controlling the hole opening of the slide gate plate. Bath depth is usually monitored by tundish weight. The meniscus level in the mold is monitored by an eddy current sensor located above the meniscus. Variation in the feed rate occurs due to the erosion of the nozzle and/or from the accretion of the alumina inclusions on the nozzle and slide plate hole. Accordingly, the feed rate is controlled by computer to offset any flow rate variation by integrating signals for the tundish weight, meniscus level, withdrawal speed, and slide plate position in the slide gate. The melt stream from the gate passes through the bore of the SEN, which is immersed in the melt pool in the mold, and exits from the side ports of SEN into the mold. The side ports, usually bifurcated for slab casting, are located near the bottom of the SEN directed toward the narrow faces of the mold. SEN ports are generally inclined at 10 to 25 degrees downward for slab casting. The number of ports for slab casting is generally two but may be up to four. The number of side ports may also be four for bloom casting. For billet casting, an open bottom SEN is commonly used.

6.13.2 *Clogging of the tundish nozzle, slide plate hole, and the SEN*

In Al-killed steel melt, fine alumina inclusions suspended in the melt are brought into contact with the bore surface (alumina) of the SEN. The turbulent diffusion of the inclusions across the velocity boundary layer developed at the SEN surface causes their contact. Upon contact, interfacial tension between the melt and the inclusions favors sticking of the inclusion particles to the bore surface of the SEN. Continued contact and attachment of inclusion particles grow the layer to form accretion to an extent that steel melt flow becomes inadequate and irregular. Accretion often occurs on the tundish nozzle, around the through hole of the slide plate, and near the bottom part of the SEN. Once an accretion develops, the melt flow is squeezed, making it impossible to cast steel melt at the predetermined throughput rate. The accretion in the SEN also causes asymmetric melt flow, which sometimes penetrates deep into the metal pool in the strand, inhibiting the flotation of inclusions [59]. Accretion can be prevented or reduced by converting alumina inclusions into liquid lime aluminates with the Ca-addition or by injecting fine bubbles into the melt path to form a gas curtain which prevents contact of alumina inclusions with the refractory surface of the nozzle bore and the slide gate through hole.

Thus, Ar injection is used in tundish nozzles and/or the SEN as shown in Fig. 6.61 by Suzuki *et al.* [59] and in Fig. 6.62 by Tsukamoto *et al.* [60]. In Fig. 6.61, tundish nozzles are shown which form uniform and fine bubbles on the upper and lower parts without causing any leakage during long periods of casting. Ar bubbles generated on the tundish nozzle are carried down to cover the surface of the SEN. On the other hand, several methods of Ar injection into the SEN are designed for achieving the same objectives as seen in Fig. 6.62. Ar injection from the tundish nozzle is more commonly practiced now than Ar injection from the SEN, at a flow rate of a few liters per min. Excessive Ar injection disturbs the steel melt/mold flux interphase, causing entrainment of mold flux as macro inclusions. Also, the excessive injection causes fine Ar bubbles to be carried with the melt flow to the solidified shell where they get entrapped. Fine alumina inclusions suspended in the melt tend to adhere to the bubbles during their travel to the shell. The entrapped Ar bubbles with attached alumina clusters result in sliver defects upon rolling and annealing of the strand since Ar does not dissolve in steel. Thus, prerequisites for reducing nozzle clogging by alumina accretion are:

(1) to reduce total oxygen content in the melt by stirring the steel melt after Al deoxidation;

(2) to minimize reoxidation of the deoxidized melt during transfer from the ladle to the tundish and the tundish to the mold by protecting the melt stream with Ar; and

(3) to minimize reoxidation of steel melt by the FeO and MnO that are contained in the ladle slag and carried over into the tundish.

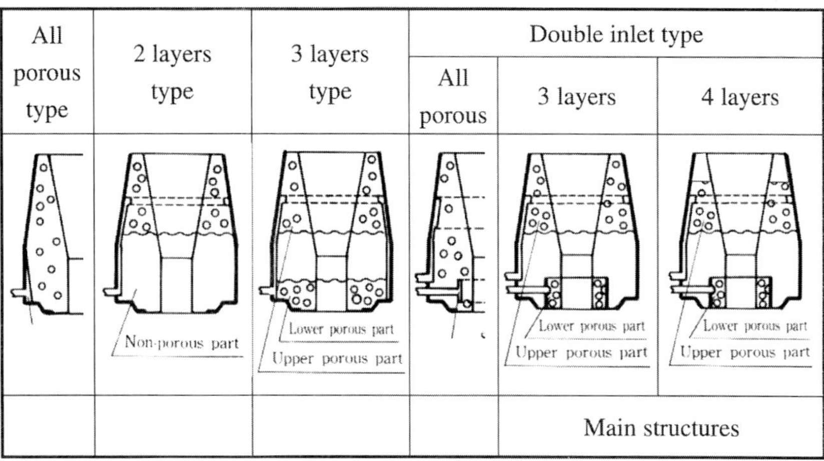

Figure 6.61: Different configurations of gas injection through tundish nozzle. [Ref. 59]

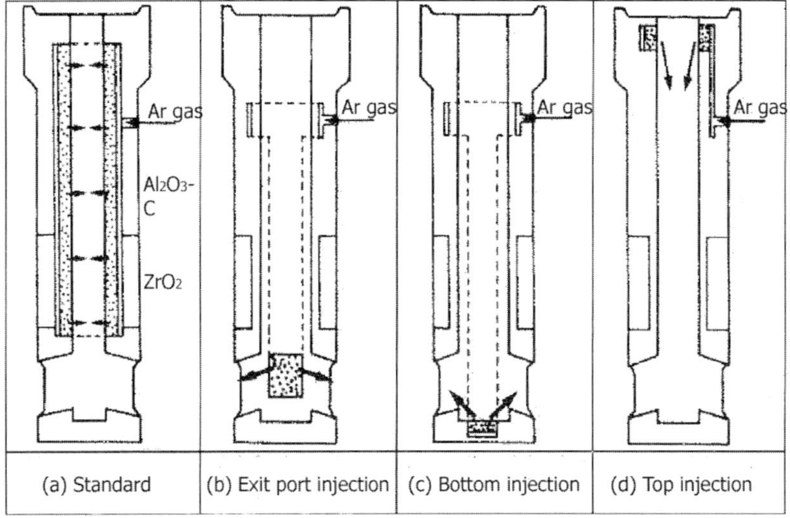

Figure 6.62: Different configurations of gas injection through SEN. [Ref. 60]

However, even taking these precautions coupled with reasonable amount of Ar injection into the nozzles have not always been found satisfactory in preventing clogging over long periods of sequential casting of many large size heats. Additional methods attempted to prevent nozzle clogging are:

(1) to use nozzles with oversize bores;
(2) to insert a CaO-ZrO_2 sleeve in the SEN bore;
(3) to insert a smooth-surfaced dense alumina sleeve in the SEN bore; and
(4) to add double annular steps on the bore of the SEN as shown by Hiroki *et al.* in Fig. 6.63 [61].

The intention of (1) is to have some extra space for metal to flow even when an accretion is formed; (2) is to let alumina react with CaO to form liquid lime aluminates; (3) is to reduce flow disturbance at the melt/sleeve boundary and also to prevent the occurrence of a chemical gradient (dissolution of carbon in an alumina graphite SEN into the melt) which attracts inclusion particles toward the bore surface; and (4) is to keep melt flow in the SEN uniform even when the slide gate is partially squeezed and to distribute Ar bubbles homogenously, as shown in Fig. 6.63. Each of these methods has met with partial success, but none has been accepted yet as a single solution by the majority of casters, probably due to the differences in the major contributing factors to the clogging at each caster.

For large scale slab casters, combined efforts to minimize all contributing factors to the clogging during the course from ladle refining to casting have made it possible to cast an acceptable number of heats in sequences. An automatic on-the-fly exchange of the SEN together with a tundish hot cycle has improved sequential casting greatly without being interrupted by the clogging. For thin slab casters and billet casters, however, the use of a thinner bore SEN is mandatory for geometrical constraints, which do not allow Ar-injection and a complicated SEN design. Accordingly, sequential casting of Al-killed steel heats is usually performed with Ca-addition to the melt to convert alumina inclusions to liquid lime aluminates that can be washed away into the mold and result in increase of macro aluminate inclusions.

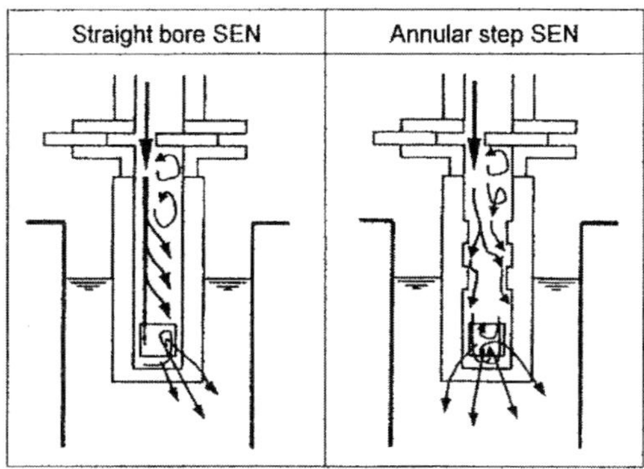

Figure 6.63: Melt flow pattern in straight bore and double annular step SEN. [Ref. 61]

6.14 Concluding Remarks

Various tundish operational and design aspects have been discussed in this chapter. Since tundish operations may differ significantly from plant to plant, every aspect may not have been covered here. Effort has nevertheless been made to incorporate the most advanced tundish technologies. Thus, for cleaner steel casting, the following aspects are considered important:

(1) A larger and deeper tundish with a covered top lid filled with argon gas atmosphere or covered by tundish flux;
(2) A ladle to tundish metal stream in a long nozzle or enclosed in a ceramic shrouding pipe with Ar gas injection;
(3) Prevention of slag transfer from the ladle to the tundish and from the tundish to the mold by slag sensing and transfer prevention technologies;

(4) A filled start to reduce macro inclusions being carried over into mold during ladle opening;

(5) High quality refractory, tundish flux, and hot tundish recycling operation;

(6) Calcium alloy addition for inclusion modification, depending on applicability when deemed necessary;

(7) Use of flow control devices depending upon the situation; and

(8) Use of an Ar injecting tundish nozzle and air-tight SEN.

Proper temperature control of metal in the tundish is another important aspect, which is covered in the next chapter.

References

1. I. D. Sommerville and E.J. McKeogh: *2nd Process Technology Conference, ISS-AIME*, 1981, 256-268.

2. H. Schrewe and F.P. Pleschiutschnigg: 2nd *Process Technology Conference, ISS-AIME*, 1981, 114-122.

3. M. Hashio, M. Tokuda, M. Kawasaki, and T. Watanabe: *2nd Process Technology Conference, ISS-AIME*, 1981, 65-73.

4. T. Ohno, T. Ohashi, H. Matsunaga, T. Hiromoto, and K. Kumai: *Trans. ISIJ*, 1975, 15, 407-416.

5. Y. Shirota, Nishiyama Kinen Kouza, NMS-ISIJ, 1990, 143-144, p. 167.

6. G.A. Demasi and R.F. Hartmann: *The Shrouding of Steel Flow for Casting and Teemimg, ISS-AIME*, 1986, 3-11.

7. N.A. McPherson: *Steelmaking Conference Proceedings, ISS-AIME*, 1985, 68, 13-25.

8. T. Nagaoka, J.P. Radot, T. Reynolds, A. Vaterlaus, and M. Wolf: *Steelmaking Conference Proceedings, ISS-AIME*, 1986, 69, 799-810.

9. Y. Tozaki, T. Hirata, A. Satoh, K. Sekino, T. Yokoi, and K. Kanda: *Steelmaking Conference Proceedings, ISS AIME*, 1993, 377-382.

10. T. Ishikura, T. Saito, T. Yasui, K. Matsuo, M. Yokoyama, and H. Fujimoto: *9th PTD Conference Proceedings, ISS AIME*, 1990, 115-121.

11. A.W. Cramb and M. Byrne: *Steelmaking Conference Proceedings, ISS-AIME*, 1984, 67, 5-13.

12. E. Julius: *Stahl u. Eisen*, 1987, 107, 397-402.

13. A. Sato, H. Nakajima, T. Sakane, T. Hamana, Y. Nakamura, and A. Oyama: *CAMP-ISIJ*, 1991, 4, 1293.

14. K. Horikawa, T. Saito, K. Enami, M. Kimura, K. Tanigawa, J. Azuma, S. Hosoya: *CAMP-ISIJ*, 1990, 3, 1212.

15. (No author listed) Sumitomo Metal Industries, *Trans. ISIJ*, 1986, 26, 590.

16. T. Itoh, T. Koshikawa, T. Imai, and A. Takahashi: *CAMP ISIJ, Part III*, 1981, 22, B-90.

17. H-J. Ehrenberg, J. Glaser, H. Jacobi, and K. Wunnenberg: *Int. Conf. Sec. Metall., Stahl u. Eisen*, 1987, 149-159.

18. R. Steffen, P. Andrzejewski, A. Diener, W. Pluschkell, M. Dubke, G. Vanino, L. Weber, D. Sucker, H. Hage-Jewasinski, and K. Schwerdtfeger: *Int. Conf. Sec. Metall., Stahl u. Eisen*, 1987, 97-118.

19. K. Matsumoto, Y. Hoshijima, K. Ishikura, K. Umezawa, Y. Nuri, and Y. Ohori: *Proceedings of the Sixth International Iron and Steel Congress, ISIJ Publication*, 1990, 3, 222-229.

20. M. Ohji, Nishiyama Kinen Kouza, NMS-ISIJ, 1994, 153-154, p. 1-39.

21. K-H. Tacke and J.C. Ludwig: *Steel Research*, 1987, 58, 262-270.

22. A.K. Sinha: *Ph.D. Thesis, The Ohio State University, USA*, 1990.

23. F. Kemeny, D.J. Harris, A. McLean, and T.R. Meadowcroft: *2nd Process Technology Conference, ISS-AIME*, 1981, 232-245.

24. D.J. Harris and J.D. Young: *Steelmaking Conference Proceedings, ISS-AIME*, 1982, 65, 3-16.

25. J. Knoepke and J. Mastervich, *Steelmaking Conference Proceedings, ISS-AIME*, 1986, 69, 777-788.

26. M. Schmidt, T.J. Rosso, and D.J. Bederka: *Steelmaking Conference Proceedings, ISS-AIME*, 1990, 73, 451-460.

27. M.L. Lowry and Y. Sahai: *Steelmaking Proceedings, I.S.S. Publication*, 1989, 72, 71-79.

28. J.D. Dorricott, L.J. Heaslip, and P.J. Hoagland: *Tundish Metallurgy, vol. II, ISS-AIME Publication*, 1991, 71-76.

29. T. Emi and Y. Habu: *Proc. Phys. Chem. and Steelmaking*, S.F.M., IRSID, ATS, 1978,126-131.

30. H. Nakajima, F. Sebo, S. Tanaka, L. Dumitru, D.J. Harris, and R.I.L. Guthrie: *Steelmaking Conference Proceedings, ISS-AIME,* 1986, 69, 705-716.

31. P. Tolve, A. Praitoni, and A. Ramacciotti: *Steelmaking Conference Proceedings, ISS-AIME*, 1986, 69, 689-697.

32. D. Bolger and K. Saylor: *Steelmaking Conference Proceedings, ISS-AIME*, 1994, 77, 225-233.

33. H. Yamanaka, T. terajima, K. Nakada, T. Koshikawa, N. Ueda, and Y. Yoshii: *Tetsu-to-Hagane*, 1983, 69, S213.

34. C. Marique, A. Dony, and P. Nyssen: *Steelmaking Conference Proceedings, ISS-AIME*, 1990, 73, 461-467.

35. T. Saeki, H. Tsubakihara, A. Kusano, K. Umezawa, and I. Suzuki: *Tetsu-to-Hagane*, 1987, 73, A207-210.

36. M. Hanmyo, M. Ishikawa, Y. Ogura, S. Matsumura, S. Miyahara, Y. Okubo: *Tetsu-to-Hagane*, 1987, 73, A215-218.

37. K. Fuchigami and M. Wakoh, *CAMP-ISIJ*, 2002, 14, 913.

38. Y. Fukuzaki, S. Kawasaki, Y. Kanazuka, T. Takebayashi, and T. Hata: *Steelmaking Conference Proceedings, ISS-AIME*, 1992, 75, 397-403.

39. N. Bessho, H. Yamasaki, T. Fujii, T. Nozaki, and S. Hiwasa: *ISIJ International*, 1992, 32, 157-163.

40. H. Uchibori: *Present and Future Perspectives of Production Techniques for Ultra Clean Steel, Nishiyama Memorial Technical Lecture, ISIJ Publication*, 1988, 126.127, 1-29.

41. H. Nakatao, K. Matsuo, M. Kimura, K. Semura, and K. Tomioka: *Tetsu-to-Hagane*, 1995,81, 709-714.

42. A. van der Heiden, P.W. van Hasselt, W.A. de Jong, F. Blaas: *Steelmaking Proceedings, ISS-AIME*, 1986,69, 755-760.

43. M. Kitamura, S. Koyama, Y. Yao, T. Soejima, and J. Anfu: *Tetsu-to-Hagane*, 1979, 65, S721.

44. Y. Yoshii, Y. Habu, T.Emi, S. Moriwaki, T. Koshikawa, and T. Imai: *Tetsu-to-Hagane*, 1978,64, S626.

45. O. Haida, T. Emi, G. Kasai, M. Naito, and S. Moriwaki: *Tetsu-to-Hagane*, 1980, 66, 48-56.

46. M. Schmidt and M.R. Ozgu: *Steelmaking Conference Proceedings, ISS-AIME*, 1986, 69, 505-510.

47. A. Diener, E. Görl, W. Pluschkell, and K.D. Sardemann: *Steel Research*, 1990, 61, 449-454.

48. M.T. Burns, J. Schade, W.A. Brown, and K.R. Minor: *10th PTD Conference Proceedings*, 1992, 75, 177-185.

49. C. Damle and Y. Sahai: *Trans. ISS, Iron & Steelmaker*, 1995, 22, No. 6, 49-59.

50. K. Taguchi, Y. Shiratani, M. Ishikawa, J. Fukumi, T. Ishida, and D. Koyanagi: *Tetsu-to-Hagane*, 1981, 67, S143.

51. K. Okuma, T. Saito, K. Enamido, K. Matsuo, A. Ote, and H. Fujimoto: *CAMP-ISIJ*, 1990, 3, 198.

52. M. Maeda, T. Saito, K. Ebato, K. Matsuo, H. Yokoyama, and H. Fujimoto: CAMP-ISIJ, 1990, 3, 199.

53. T. Shirai: *Shin Nittetsu (Nippon Steel Technical Report)*, 1993, 351, 21-26.

54. K. Isono, T. Shirai, T.Hiraoka, T. Ohno, K. Ohnuki, A. Watanabe, K. Fujii, and H. Kasahara: *Steelmaking Conference Proceedings, ISS-AIME*, 1994, 77, 249-253.

55. H. Tanaka, R. Nishihara, R. Miura, R. Tsujino, T. Kimura, T. Nishi, and T. Imoto: *ISIJ International*, 1994, 34, 868-875.

56. Y. Tozaki: *Recent Trends in Quality Control Technology of Continuously Cast Slabs, Nishiyama Memorial Technical Lecture, ISIJ Publication*, 1994, 153/154, 159-187.

57. K. Kanda, K. Sekino, A. Sato, T. Sakane, T. Yokoi, and N. Yoshida: *CAMP-ISIJ*, 1992, 5, 313.

58. Tekko Binran (Handbook of Iron and Steel) 3rd Ed., Vol. II, *Ironmaking and Steelmaking, ISIJ*, Maruzen, Tokyo, 1981, p. 635.

59. H. Suzuki, Y. Yoshimura, N. Ogata, and M. Imai: *Shinagawa Tech. Report*, 2003, 46, 67-76.

60. N. Tsukamoto, K. Kanamaru, E. Iida and K. Yanagawa: *Shinagawa Tech. Report*, 1993, 36, 3, 89-98.

61. N. Hiroki, A. Takahashi, Y. Namba, N. Tsukamoto, Y. Kurashina and K. Yanagawa: Shinagawa Technical Report, 1993, 36, 75-88.

Chapter 7

Melt Temperature Control

7.1 Introduction

Metal temperature in a tundish has a strong influence on the quality and properties of cast product, caster operation, and refractory life. A large ladle may pour metal into the tundish for as long as one hour at the caster. Temperature losses are always inherent in the system, specifically in the ladle and by the tundish from the melt surface and through the refractory walls. Thus, the temperature of the ladle-to-tundish stream continuously changes. Coupled with the temperature loss in the tundish, temperature changes in the stream result in melt temperature variation within the tundish. Since superheat control is a very important parameter for controlling the cast structure and quality of a metal, external heating of melt in the tundish is now becoming a more common operation in plants where the quality of cast metal is an important consideration. This chapter first describes the temperature variation in a tundish, and then discusses the external heating and cooling methods employed in industry.

7.2 Melt Temperature Variation

Several continuous melt temperature measurements have been made in industrial tundishes [e.g. Refs. 1, 2]. Figure 7.1(a and b) [1] shows the change of metal temperature in a tundish for two different heats. Figure 7.1a shows an abnormally large temperature variation of about 30 °C during one ladle, whereas Fig. 7.1b represents a more normal operation with a temperature drop of about 15 °C. Figure 7.2(upper) [2] is a result of continuous temperature measurement at the Bethlehem Steel

Corporation. This also shows the classical dome-shaped temperature profile with a variation of about 20 °C. Figure 7.2(lower) [2] shows the tundish melt temperature variation for casting of 4 ladles. 'A' in this figure represents the ladle change, and points to the fact that each ladle may start with a different temperature, and that the highest and lowest melt temperatures in a tundish may vary from ladle to ladle. The melt temperature in the tundish in each of these cases first rises and reaches a maximum value in about 20 minutes, followed by a gradual decrease for the rest of the heat. Thus, at the end of a ladle, metal in the tundish is at its lowest temperature. The ladle is then changed, and a new ladle with a significantly higher melt temperature is brought in. Metal is poured into the tundish at a flow rate higher than the steady state rate to bring the metal level to the normal operational depth. This causes the temperature of metal in the tundish to rise to its highest value. During this period, the temperature of the ladle stream continues to drop, and at the same time, the tundish also loses heat from the walls and free surfaces. A continuous gradual temperature drop in the tundish, after reaching the highest temperature, results from these heat losses in the ladle and the tundish.

Chakraborty and Sahai [3] developed mathematical models for coupling heat transfer and fluid flow in a typical ladle and a slab casting tundish over a complete casting sequence. The complete casting sequence, as considered in their study, consisted of one minute period of ladle change, a one-minute filling period with a new ladle to bring the metal level to a normal depth, and a 46-minute period of steady state casting. In one extreme case, Chakraborty and Sahai considered a large heat loss from the ladle surface by assuming an uncovered ladle. Curve A in Fig. 7.3 shows the predicted ladle stream temperature drop during 48 minutes of casting. The stream temperature drops by about 40 °C in this period. The rate of stream temperature drop is smaller in the beginning but increases towards the latter part of the casting. This is due to the smaller volume of metal in the ladle which suffers a relatively higher heat loss towards the end of the casting period. Curves B, C, and D represent predicted melt temperatures in the tundish at 3 different monitoring points. The melt temperature is predicted to rise to a maximum value in about 25 minutes, and is followed by a gradual decline. A well preheated, well covered, and well insulated ladle will have less stream temperature loss, and in turn, will affect the tundish melt temperature.

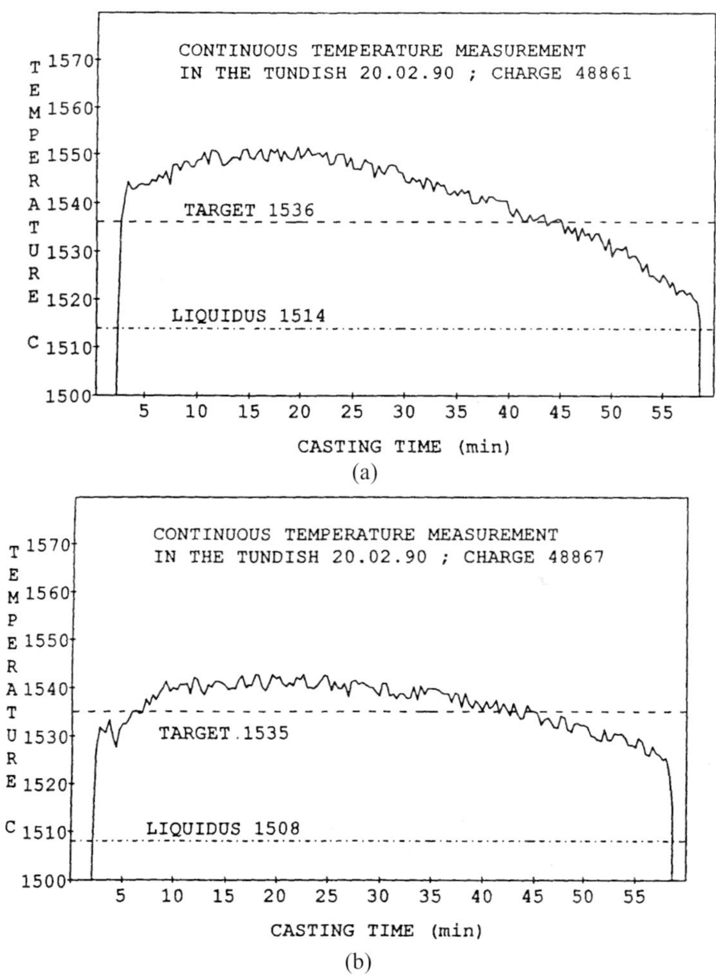

Figure 7.1: (a and b) Tundish melt temperature variation for two heats. [Ref. 1]

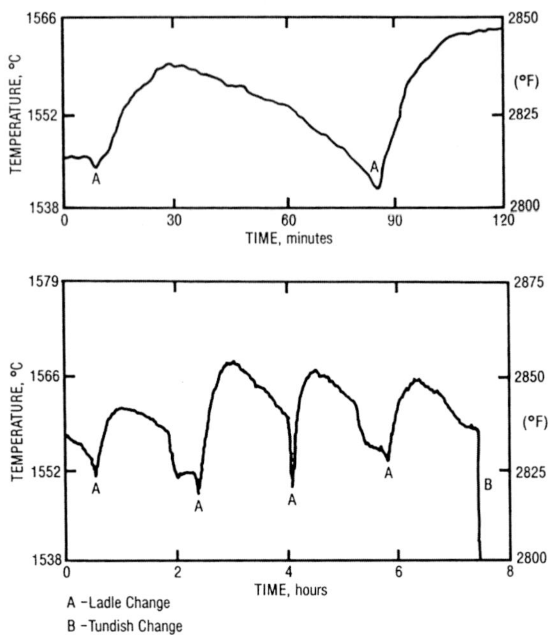

Figure 7.2: Tundish metal temperature profile during normal casting sequence. [Ref. 2]

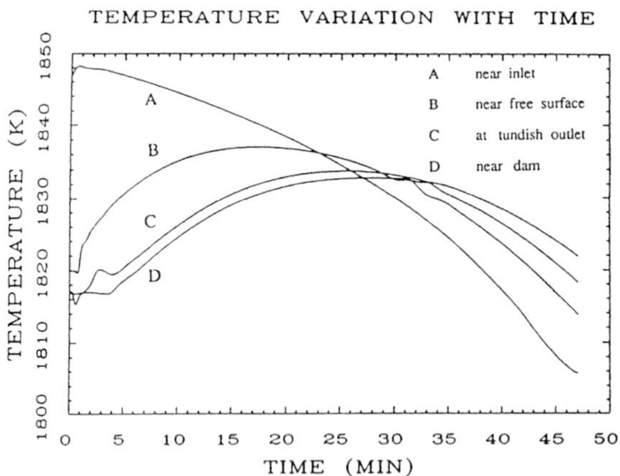

Figure 7.3: Metal temperature variation in a tundish at four monitoring points. [Ref. 3]

Similar conclusions were reached by Miller and Hlinka [4] who performed a water modeling study in a Plexiglas ladle. These researchers used boiling water to simulate molten steel. The experiments were carried out with two different thicknesses of mineral oil on top of the boiling water. This water modeling study monitored the variation of the fluid temperature during teeming. Figure 7.4 represents the variation of water temperature during teeming for three different cases, the times and temperatures having been scaled to correspond to a 100-ton heat. Case 1 corresponds to a thin layer of oil (normal slag) where the temperature drops continuously due to heat losses from the surface and walls. Case 2 corresponds to a situation where the thick layer of oil was used but the bulk fluid was stirred during teeming, while case 3 corresponds to normal teeming from the ladle with a thick oil layer. Case 1 has the maximum temperature drop, and case 3 has the lowest. The mixing in case 2 causes the hotter fluid from the relatively stagnant central bulk to move near the walls, and thus, loses more heat than case 3.

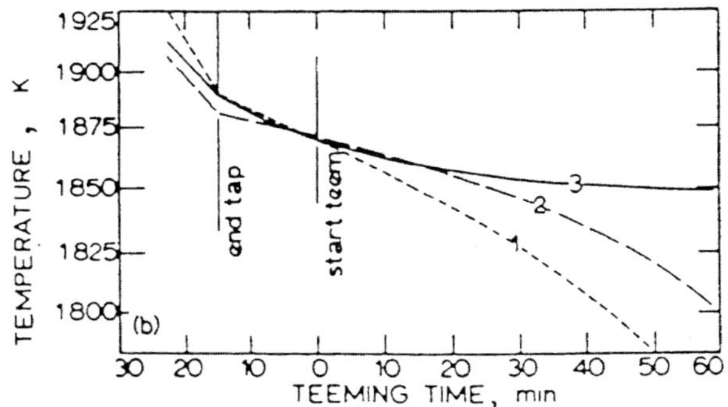

Figure 7.4: Melt temperature variation during teeming from a ladle as predicted by a water model study. [Ref. 4]

Chakraborty and Sahai [3] found that the flow and heat transfer characteristics of melt in the tundish operating at normal melt depth after the ladle transfer operation were related to the temperature variation of the incoming ladle melt stream. If the temperature of the ladle stream exhibited a significant drop over the casting period, as is shown in Fig.

7.3, the stream became cooler than the bulk temperature in the tundish after about 30 minutes of teeming. This resulted in a bottom-directed flow of the melt instead of the normal top surface-directed flow. The normal flow is shown in Fig. 7.5a, and the bottom directed inverse flow is shown in Fig. 7.5b. This inverse flow pattern persisted for the remainder of the 47-minute casting period, but reverted back to the free surface directed flow during the one-minute filling period with the subsequent heat at a higher temperature.

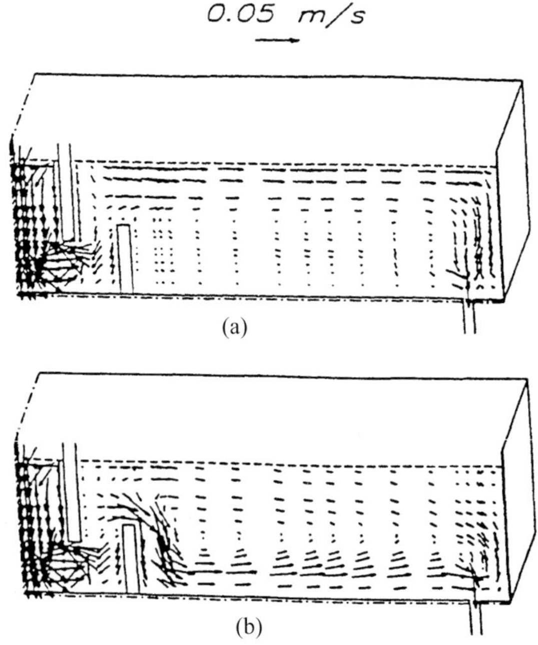

Figure 7.5: (a) Metal flow field in a tundish under steady-state casting, and (b) effect of ladle stream temperature variation on tundish flow field. [Ref. 3]

7.3 Temperature Measurement Devices

Tundish temperature measurement is accomplished by the use of thermocouples, which may be disposable or intended for continuous temperature measurement. The disposable system has a quick response, generally taking about five seconds, and has high degree of reliability. Continuously-measuring thermocouples can work for a relatively long time. Russo and Phillippi [2] used the type B (Pt - 6% Rh / Pt - 30% Rh) thermocouple in an enclosed ceramic insulator. The assembly was then placed in a closed-end molybdenum sleeve. The two-piece system is shown in Fig. 7.6. A carbon steel extension pipe was used to extend the thermocouple leads to a cool zone. The system was then enclosed in an alumina-graphite iso-pressed composite, which was able to survive a full tundish campaign of 10 to 12 hours. The thermocouple leads were connected to a high temperature copper lead with an armor jacket, and the joint was cooled by compressed air. At operating temperatures, the thermal lag, to within several degrees C of the true temperature, was about 90 sec. The authors claim that the system worked very well, with the average life of the thermocouple assembly exceeding 100 hours.

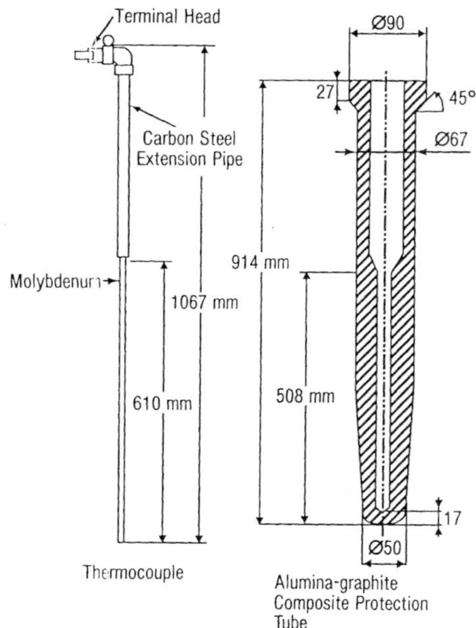

Figure 7.6: Two piece temperature measurement system, thermocouple and ceramic insulator. [Ref. 2]

Mabuchi *et al.* [5] at Kawasaki Steel Corporation used a type R thermocouple enclosed in Mo-cermet (Mo: 60%, ZrO_2: 40%). This was then enclosed in a zirconia-graphite refractory. The system is schematically shown in Fig 7.7. Mori *et al.* [6] of Nippon Steel Corporation developed a continuous temperature measuring system which is schematically shown in Fig. 7.8. They used a Pt/Rh thermocouple enclosed in ZrB_2, which was then sealed in an Al_2O_3 tube. The temperatures measured with this continuously-measuring thermocouple were compared with those measured by the conventional disposable thermocouple. The results are plotted in Fig. 7.9, and the system is found to be very accurate. It accurately measured temperature continuously at Nippon Steel for an average of 40 hours and to a maximum of more than 100 hours during carbon steel casting.

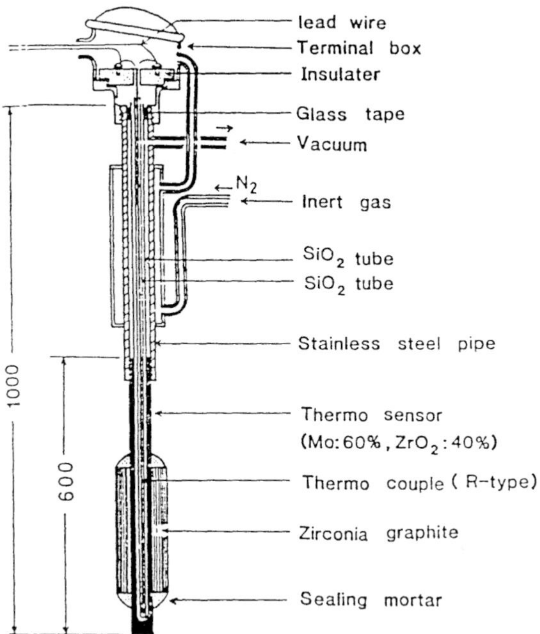

Figure 7.7: Thermocouple and Mo cermet enclosed in a zirconia-graphite refractory for continuous measurements. [Ref. 5]

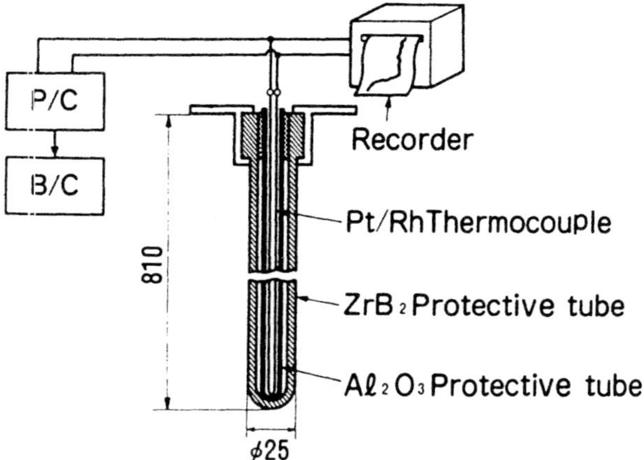

Figure 7.8: Pt/Rh thermocouple enclosed in ZrB_2, which was sealed in an alumina tube. [Ref. 6]

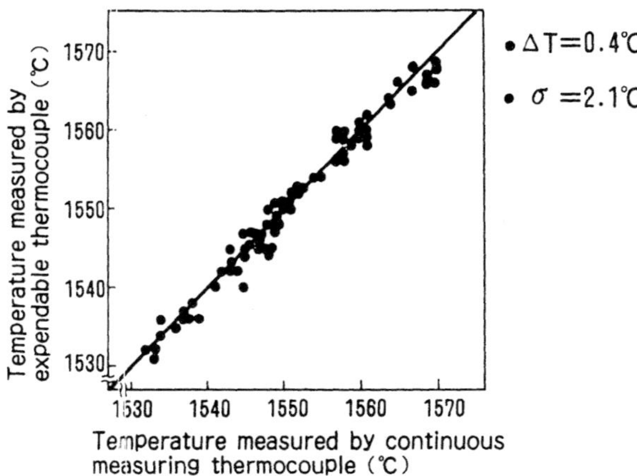

Figure 7.9: Temperature measured by the thermocouple (in Fig. 7.8) compared with those measured by a disposable thermocouple. [Ref. 6]

7.4 The Need for Superheat Control

The quality and internal structure of cast product are largely dependent on the amount of superheat in the melt. The relative extent to which equiaxed and columnar structures form in the solidified steel depends on many variables, including degree of superheat, steel chemistry, and section size. Takeo *et al.* [7] studied the effect of melt superheat on the area of equiaxed crystals in 0.8 %C steel continuously cast into a 110 mm square billet at 2.4 m.min^{-1} casting speed. Their results are plotted in Fig. 7.10, which shows that the relative area of the equiaxed zone increases as superheat decreases. At higher melt temperatures, crystals that have started to form are re-melted, and very few are left to grow into equiaxed crystals. Thus, the long columnar dendrites grow to the billet centerline. The columnar structure causes segregation, internal cracks, and porosity.

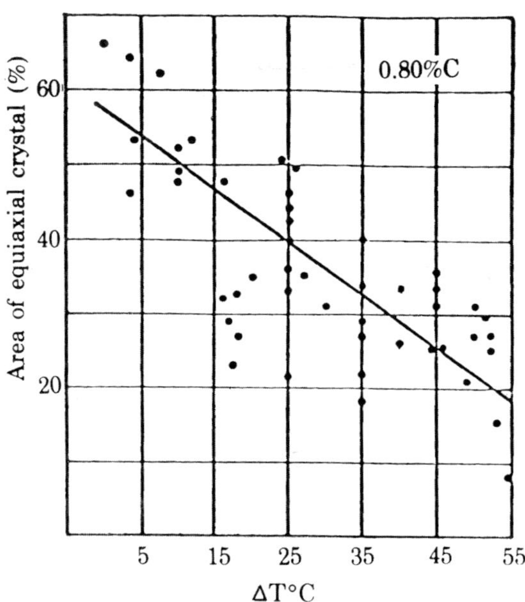

Figure 7.10: The effect of melt superheat on equiaxed crystals. [Ref. 7]

In terms of quality, the number of non-metallic inclusions, including slag entrapment, sharply increases with decreasing temperature. Figure 7.11 [8] shows the effect of tundish steel temperature on the inclusion index and centerline segregation index in a billet. Matsumoto *et al.* [9] also studied the influence of tundish melt superheat on the index of inclusion in cast slab. Their results are shown in Fig. 7.12, which demonstrates that the inclusions increase at low as well as high superheat. At low superheat, globular steel crystals form near the meniscus, which descend down toward the melt pool end. During this descend they capture and drag down the ascending inclusions to the bottom of the pool. High temperatures promote the reactivity of molten metal with refractory and molten slag, and result in a more contaminated product.

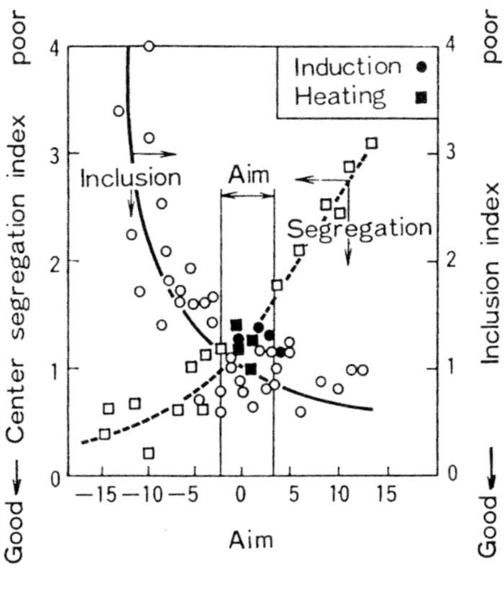

Figure 7.11: The effect of melt temperature on the inclusion index and centerline segregation index. [Ref. 8]

In addition to influencing the quality and structure of metal, the superheat also affects the operational parameters of the caster. At low temperatures, molten metal may freeze in the submerged entry nozzle, or alumina inclusions may deposit in the SEN and eventually block the nozzle. High temperature also decreases the refractory life in the tundish. Thus, the discussion clearly points to a strong need for maintaining the melt temperature within a specified narrow range. The usual target for a superheat in billet casting is 15 °C, and in slab casting the goal is 25 °C.

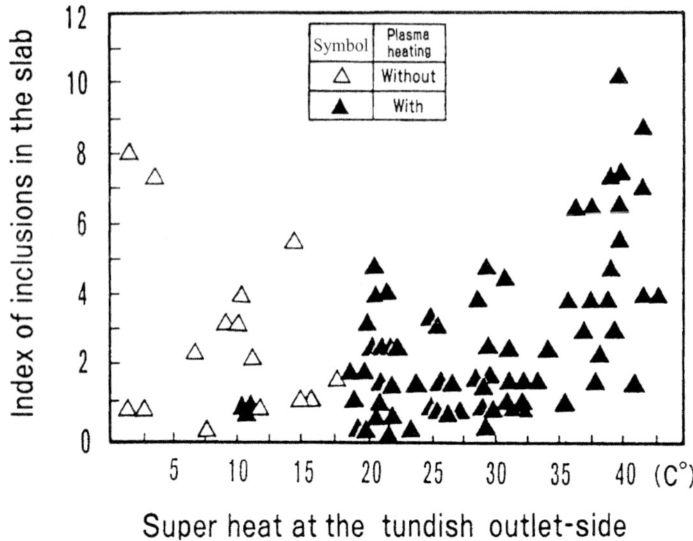

Figure 7.12: The influence of tundish melt superheat on the index of inclusion in cast slab. [Ref. 9]

Since melt superheat temperature in the tundish varies from 15 to 30 °C during the pouring of one ladle, the BOF tap temperature has to be kept high enough to allow the melt temperature not to fall below the prescribed limit for the quality of cast product and caster operation. With heating of metal in the tundish and its ability to keep temperature within a very narrow range, the BOF tap temperature can be reduced by as much as 25 °C. This increases refractory life and metal yield. Figure 7.13 [9] schematically illustrates the temperature changes from BOF tapping to tundish operation, and shows the effectiveness of tundish heating in reducing the BOF tap temperature. Finally, Fig. 7.14 [9] reemphasizes the need for a tundish heater.

Many companies have developed heating of metal in a tundish by plasma or induction, and they have been able to maintain the melt temperatures within a very narrow range, ± 1 [Ref. 10] to ± 5 °C [Ref. 9] of the target temperature.

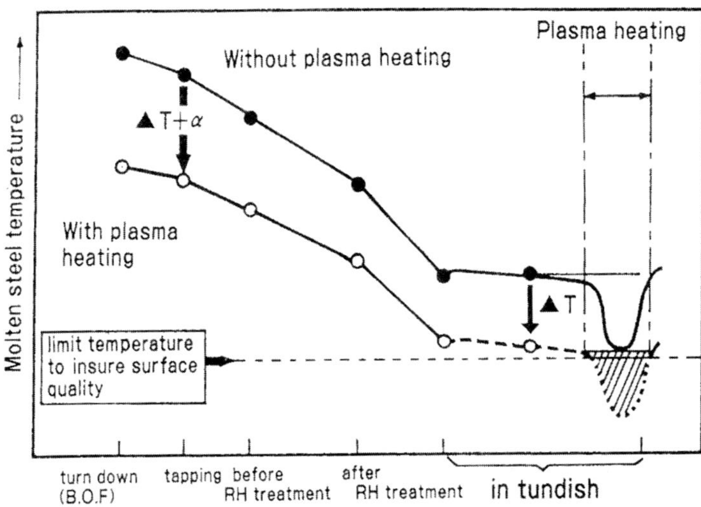

Figure 7.13: Temperature change from the BOF to tundish and effect of plasma heating. [Ref. 9]

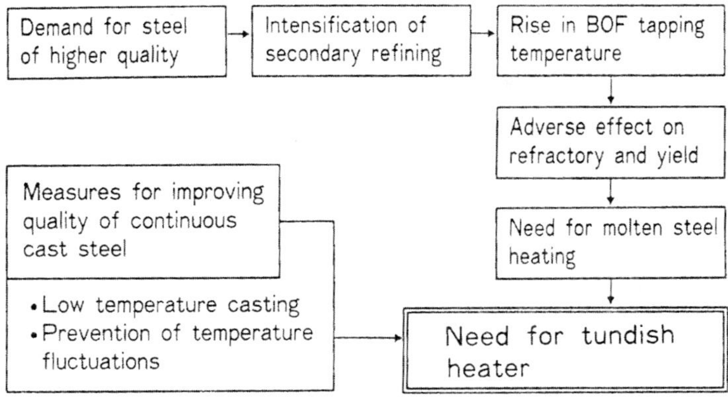

Figure 7.14: Need for tundish melt heating and control by plasma. [Ref. 9]

7.5 External Heating (Plasma Systems)

Two types of plasma heating systems are employed in tundishes. They are the DC arc and AC arc plasma systems.

7.5.1 *DC arc plasma system*

In this system a plasma torch, which acts as a cathode, is lowered from the top of an enclosed refractory lined chamber in a tundish. The torch strikes an arc with the molten metal, and to complete the electrical circuit, an anode is installed in contact with the molten metal to make the metal bath act as an anode. The torch (cathode) and the anode are connected to a DC power supply. A schematic diagram of the DC plasma heater in a tundish is shown in Fig. 7.15 [9]. A plasma torch is schematically shown in Fig. 7.16 [10]. The torch is made of a water cooled copper tube with a thoriated tungsten tip. The average life expectancy of the tungsten tip is approximately 100 hours of operation [11]. A de-ionized water supply is used for cooling the torch to avoid any mineral deposits inside the torch.

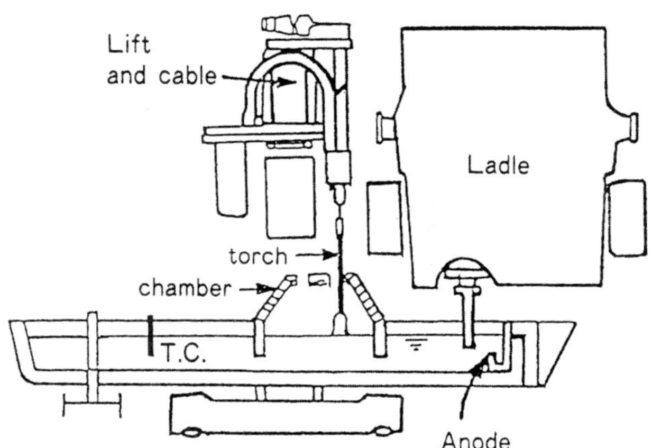

Figure 7.15: Schematic diagram of the DC plasma heater in a tundish. [Ref. 9]

The torch needs an ionization gas to form the plasma arc. The potential gradients of various gases relative to air are given in Table 7.1 [10]. Many of these gases, with a large potential gradient, are not suitable for use in the plasma torch as they oxidize the tungsten tip of the torch, deteriorating the metal quality. Such gases raise safety concerns (e.g. CO_2), or are not economically viable (e.g. He). Ar and N_2 are the only two potential candidates. Fig. 7.17 shows the relationship between the arc length and arc voltage for Ar and N_2 gas mixtures. Arc length increases with arc voltage, and the amount of N_2 in the gas mixture increases arc voltage. For the same current, as arc voltage increases, the plasma power increases, and thus, results in a higher rate of heating. The use of N_2 in the gas mixture is beneficial for getting more plasma power; however, as shown in Fig 7.18, it leads to nitrogen pick-up in the metal. These results were obtained at the NKK's experimental 200 kW (1kA x 200 V) DC plasma torch facility for heating metal in a tundish [10]. Thus, depending upon the grade of steel being cast, N_2 gas may be used in the mixture.

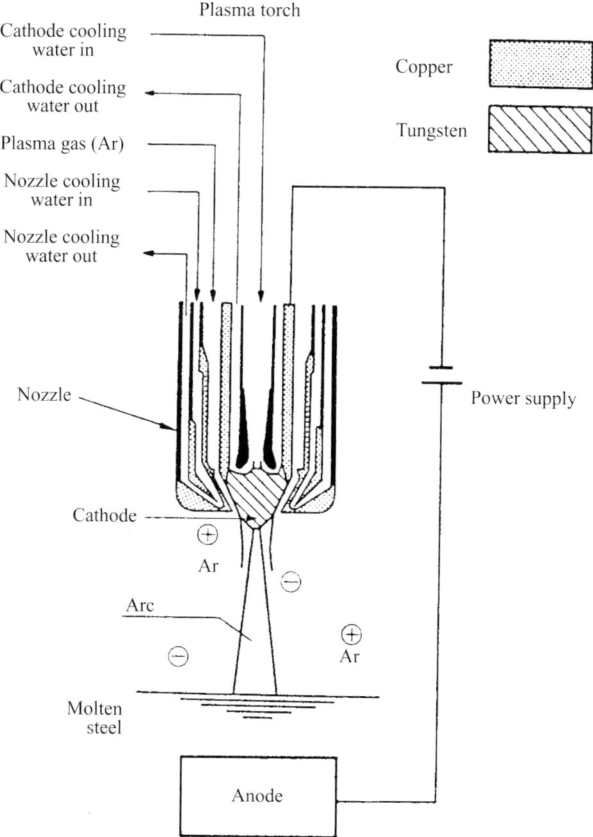

Figure 7.16: Schematic of a DC plasma torch. [Ref. 10]

Table 7.1: Potential Gradients of various gases. [9]

Gas	Index of voltage gradient	Problem	Result
Ar	0.5	-	Yes
N_2	1.1	[N] content	Maybe
Air	1.0	[N], [O]	No
CO_2	1.5	[O], Safety	No
O_2	2.0	[O]	No
H_2O	4.0	[H], [O]	No
H_2	10	[H]	No
He	1.5	High cost	No

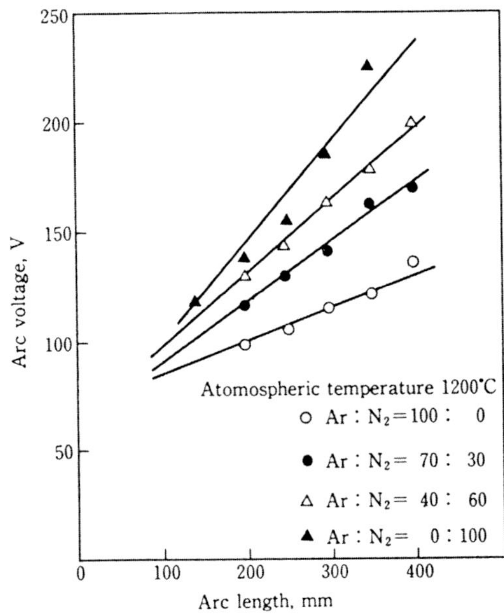

Figure 7.17: Relationship between the arc length and arc voltage for Ar and N_2 gas mixtures. [Ref. 10]

Heating efficiency - Figure 7.19 [9] shows the DC plasma arc modes of heating and various heat losses in the system. About 9% of heat is lost to the cooling water, and about the same amount is lost through the refractory of the torch enclosure, where the temperature reaches about 1800 °C. The only other loss is the approximately 2% associated with the exhaust gas. Thus, about 80% of heat is transferred to the metal, and half of this heat comes as radiation from the refractory. The other half is either from direct radiation of the plasma arc or as direct voltage drop in the metal bath. Regular operation results of Matsumoto *et al.* [9] show that a heating efficiency in the range of 60 to 80% is achieved.

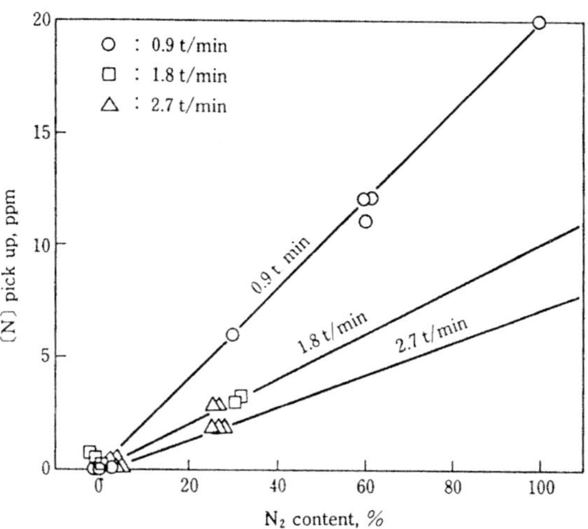

Figure 7.18: Effect of nitrogen content on the nitrogen pick-up. [Ref. 10]

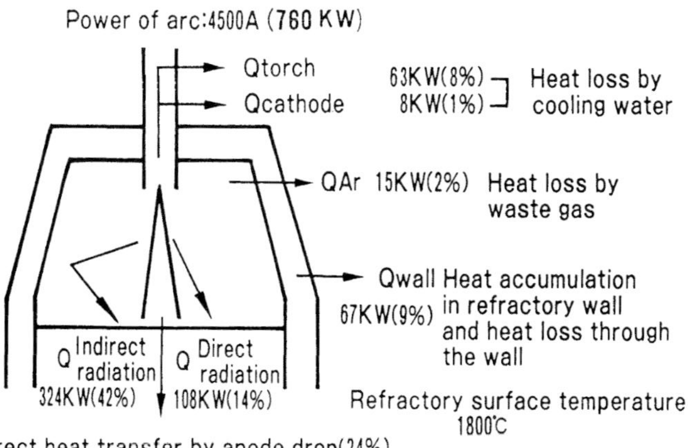

Figure 7.19: Heat losses and thermal efficiency in DC plasma heating unit. [Ref. 9]

Stirring in metal below the plasma torch helps in bringing fresh metal to the surface, and improves the heating efficiency. Tanaka *et al.* [10] tried gas stirring or the use of a dam for stirring metal in the heating chamber, which is schematically shown in Fig. 7.20. They found that both systems were equally effective in increasing the heating efficiency. In their regular operation with 1.4 MW plasma, a heating efficiency of 75 % and a torch life of over 100 hours have been achieved. They have been able to control temperature within ± 1 °C during steady-state operation, and allowed no more than 5 °C temperature drop during ladle change operation.

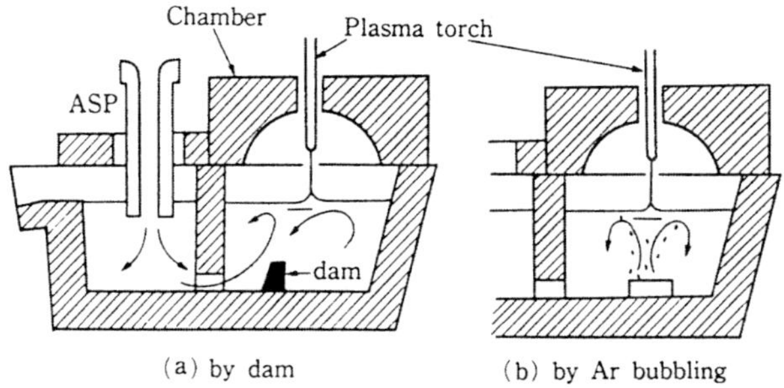

Figure 7.20: Melt stirring during plasma heating by (a) a dam and (b) gas injection. [Ref. 10]

7.5.2 *Mobile plasma arc system*

As stated in the preceding section, quality considerations demand the use of Ar gas in a DC plasma torch. However, the arc power, without compromising the quality, can only be increased by either increasing the current or the voltage of the plasma arc. Increasing the current of the system increases the wear and consumption of the tungsten cathode tip. The voltage of the system can be increased by increasing the distance of the cathode from the melt surface. But increasing the distance causes a decline in the heating efficiency. Recently, Inokuchi *et al.* at NKK [12] found that the plasma arc voltage can be doubled by making a controlled mobile arc, which is described below.

7.5.2.1 *Stable and mobile arc*

A stable DC arc can be created between a pointed carbon cathode and a flat carbon black anode. A section of such a cathode is shown in Fig. 7.21, and the arc created by such a cathode, shown in Fig. 7.22a, may be termed as a stable arc. After some use of the cathode, when its edges get rounded and the thickness of the tip exceeds 4 mm, the arc becomes unstable. Such an arc is shown in Fig. 7.22b, and is called a mobile arc. The drop in the arc column voltage of this mobile arc is about twice as large as that of the stable arc, as shown in Fig. 7.23. Thus, generation of a controlled mobile plasma arc could substantially increase its voltage for the same gap, and hence more power could be achieved.

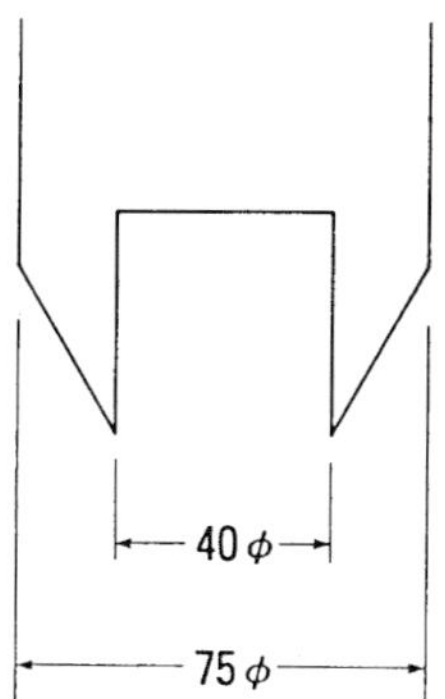

Figure 7.21: Dimensions of carbon electrode used for generating stable arc. [Ref. 12]

7.5.2.2 *Mobile plasma arc generation*

Inokuchi *et al.* [12] developed a mobile transferred arc DC plasma by two methods, which are shown in Fig. 7.24 and are described below. The end of the tungsten cathode was given hemispherical shape to provide a spread to the plasma arc.

(1) The first method was the injection of a gas from a side tube to disturb the plasma arc. The gas injection was by one or more tubes, which were bent towards the axis of the cathode or bent circumferentially to generate a rotational plasma arc about the axis of the plasma torch.

(2) The magnetically operated plasma torch nozzle was enclosed in a coil. When a current is passed through the coil, a magnetic field is set up in the plasma field. This rotating electromagnetic force tends to rotate the plasma arc. If the axis of symmetry of the current path does not coincide with the axis of the plasma arc, then a whirling plasma is formed due to superimposition of a kink instability. The resulting plasma is similar to the one shown in Fig. 7.22b.

The voltage drop in a plasma arc, for a given gap, was measured for stable and mobile plasma, which is plotted in Fig. 7.25. There was a significant increase in arc voltage for the same gap. Thus, using mobile plasma can improve the effectiveness of tundish heating.

 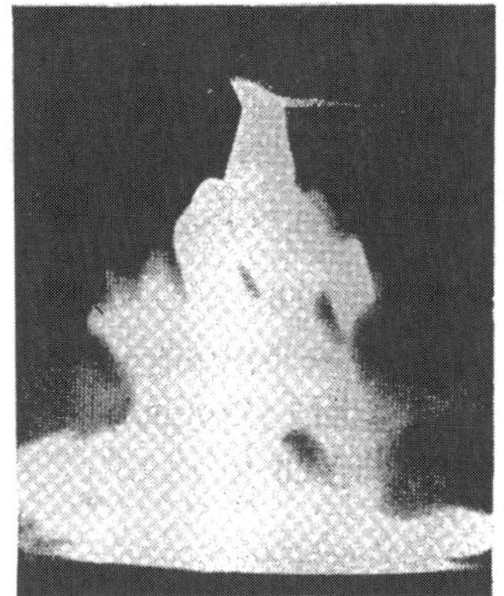

(a) Stable carbon
electrode arc
(DC 600A, 150mm)

(b) Mobile carbon
electrode arc
(DC 900A, 150mm)

Figure 7.22: Typical (a) stable plasma arc, and (b) mobile arc on a carbon electrode. [Ref. 12]

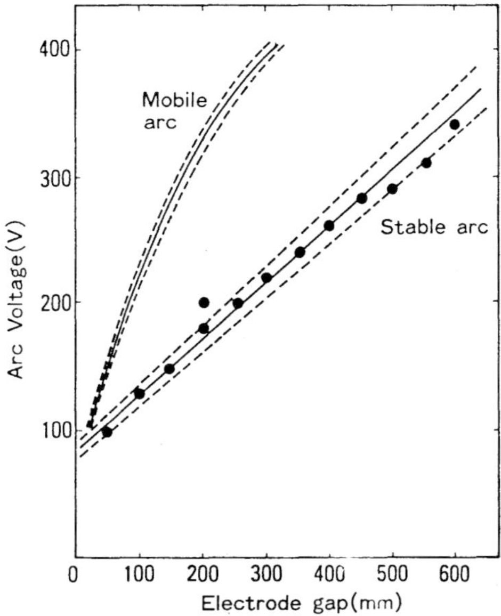

Figure 7.23: Comparison of the arc voltage between stable carbon electrode arc and a mobile arc. [Ref. 12]

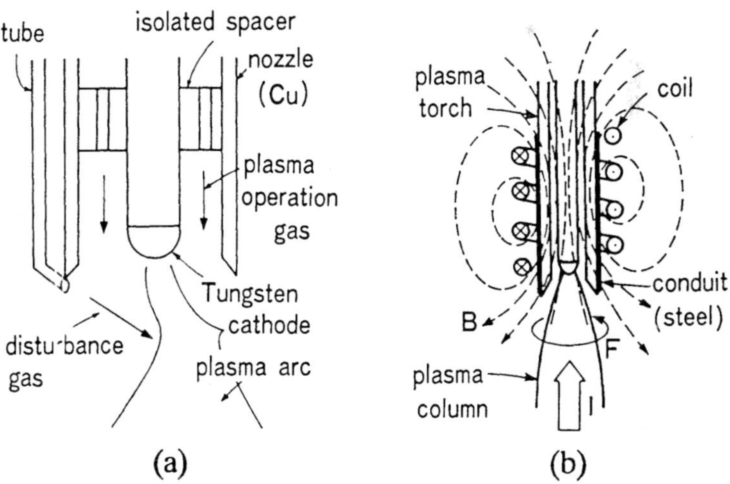

Figure 7.24: Mobile plasma torch temperating by (a) gas injection, (b) magnetic field. [Ref. 12]

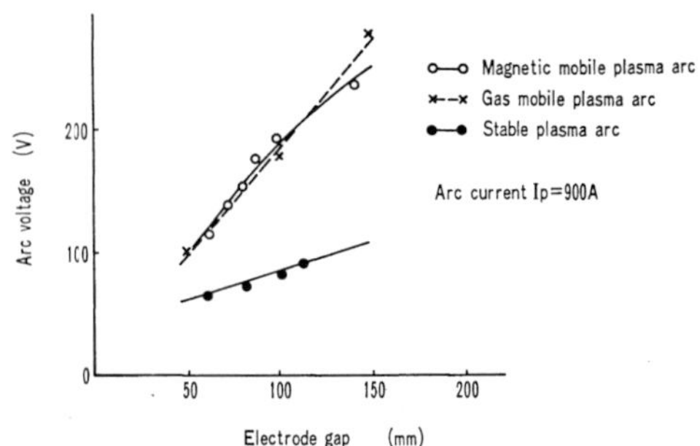

Figure 7.25: Comparison of the arc voltage between mobile arc plasma and stable plasma arc. [Ref. 12]

7.5.3 *AC arc plasma system*

Most of the description presented in this section has been compiled from Kobe Steel's publications [13-16]. Figure 7.26 schematically shows the top and side views of an AC plasma system installed in Kobe Steel's tundish. The heating system consists of two plasma torches inserted from the tundish cover, inclined at an angle, above the steel surface. The installation is in an 80-ton tundish with a metal depth of 2 m. Ar gas is injected from the bottom of the tundish in the heating zone to provide mixing and to renew steel at the free surface. The maximum power of the plasma system is 2.4 MW. Specifications of the system are provided in Table 7.2.

Table 7.2: Specifications of the AC plasma system.

Item	Specification
Type	Transferred Arc / Non-Transferred Arc
Torch diameter	108 mm
Working length	4.2 meters (approximately)
Type of current	1 phase AC
Arc current	7.5 kA maximum
Arc voltage	350 V maximum
Power	2.4 MW
Argon gas	9 to 20 Nm^3/h (per torch)
Cooling water	30 m^3/h

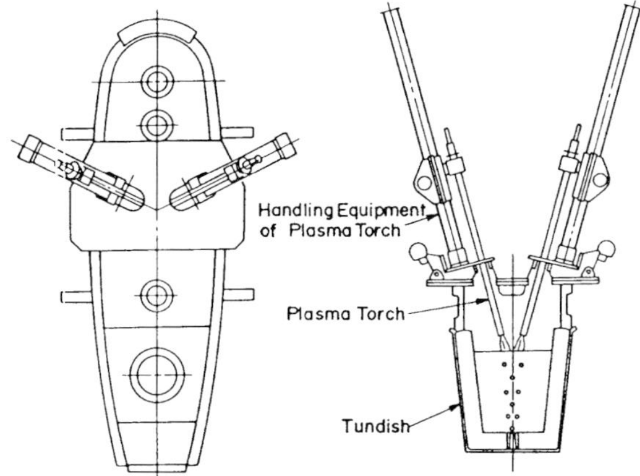

Figure 7.26: Top and side views of a single phase AC plasma. [Ref. 13]

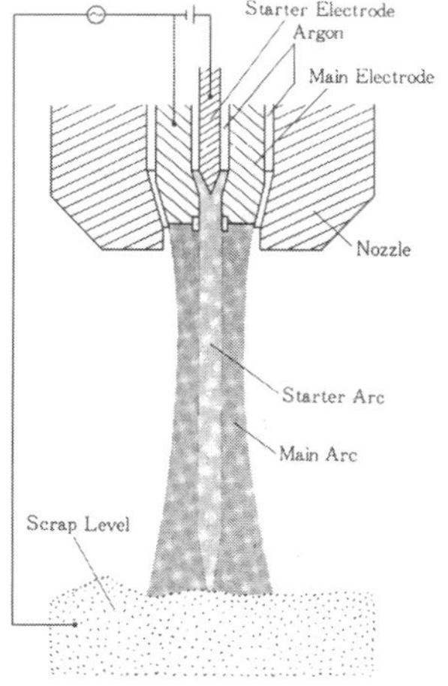

Figure 7.27: Basic design of AC transferred arc plasma torch. [Ref. 15]

In the transferred arc mode, the arc forms between each torch and metal surface, and in the non-transferred arc mode, the arc is established between the two torches. This is achieved by varying the distance between the torches. The basic design of the torch is schematically shown in Fig. 7.27. Argon gas flows through the torch, and the starter and main electrodes ignite the torch. The arc is then transferred to the metal, and electricity flows between the main electrode and the metal.

7.5.3.1 *Arc characteristics*

Figure 7.28 shows the relationship between the arc current and voltage for different arc lengths. An increase in the arc current does not change the arc voltage, but an increase in the arc length increases its voltage.

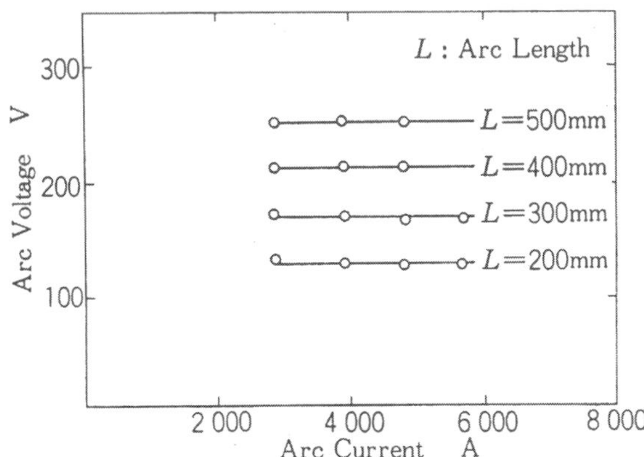

Figure 7.28: Relationship between the arc current and voltage for different arc lengths. [Ref. 13]

7.5.3.2 *Effect of gas composition and flow rate*

The presence of diatomic gases, like O_2 or N_2, in the tundish increases the arc voltage. A diatomic gas requires dissociation energy, and increases the energy required for its ionization. Thus, the frequency of collision among electrons increases, which results in an increased potential gradient. Figures 7.29 and 7.30 show the effects of the presence of different amounts in the Ar gas of O_2 and N_2, respectively, on the arc voltage. As discussed in the DC plasma section, increasing the N_2 content of the plasma gas is one way of increasing arc voltage, and hence, the power of the plasma. Depending upon the steel grade, N_2 enrichment may be one method of increasing plasma power. The influence of O_2 and N_2 content of the gas in increasing the arc voltage is shown in Fig. 7.31. The tundish was sealed, and an inert atmosphere was maintained by the flow of Ar gas in the tundish. The effects of the Ar gas flow rate and the presence of N_2 in the sealing gas are shown in Fig. 7.32. Arc voltage decreases with increasing gas flow rate, but increases with the presence of N_2 gas.

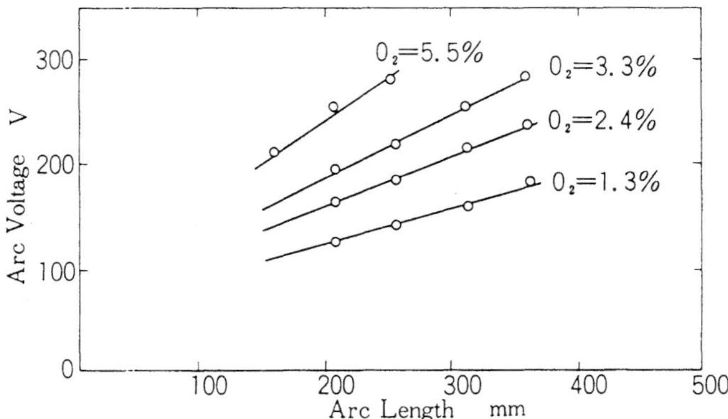

Figure 7.29: Effect of oxygen in gas on the arc voltage and length. [Ref. 13]

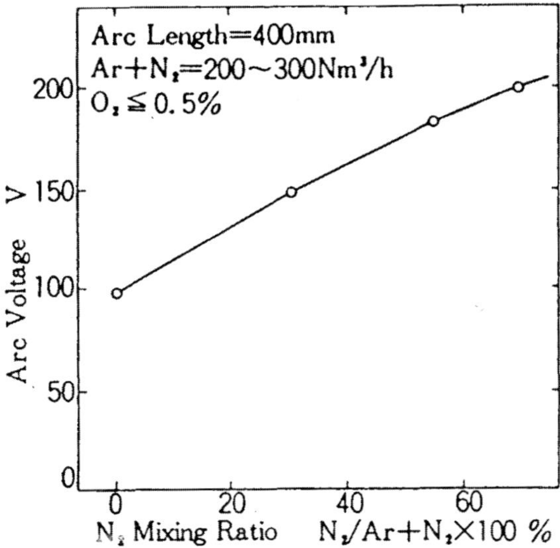

Figure 7.30: Effect of nitrogen in Ar gas on the arc voltage. [Ref. 13]

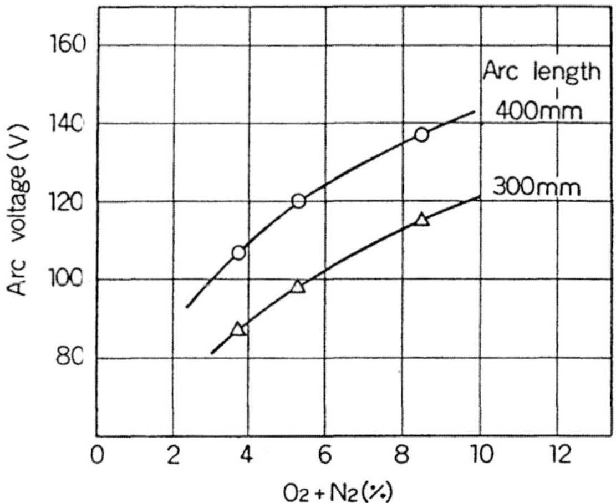

Figure 7.31: The influence of O_2 and N_2 content of the gas in increasing the arc voltage. [Ref. 13]

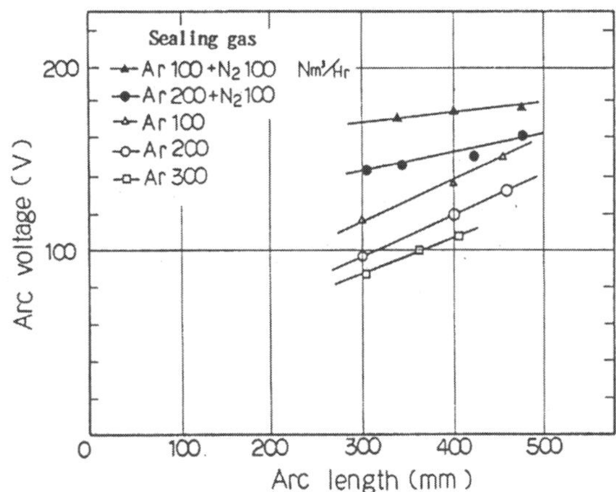

Figure 7.32: The effects of the Ar gas flow rate and the presence of N_2 in the sealing gas on arc voltage. [Ref. 13]

7.5.3.3 *Cold- and hot-run trials*

Experiments were conducted by generating an AC plasma arc on a carbon block and on molten steel. Arc voltages for different arc lengths for these two systems are shown in Fig. 7.33. The voltage level in molten steel is about 100 to 130 V under an Ar atmosphere containing less than 0.5 % oxygen, which is much less than levels achieved in carbon block trials. This difference was attributed to the presence of metal fumes in the plasma arc. The presence of metallic vapor decreases the resistance of the plasma arc, and consequently, low voltage is attained. This conclusion was confirmed by experiments performed with a thin (~ 5 mm) and thick (> 20 mm) slag cover over the melt. As shown in Fig. 7.34, voltage was about 20 V higher with the thick slag cover. Thick slag reduced the amount of metallic vapor in the gas.

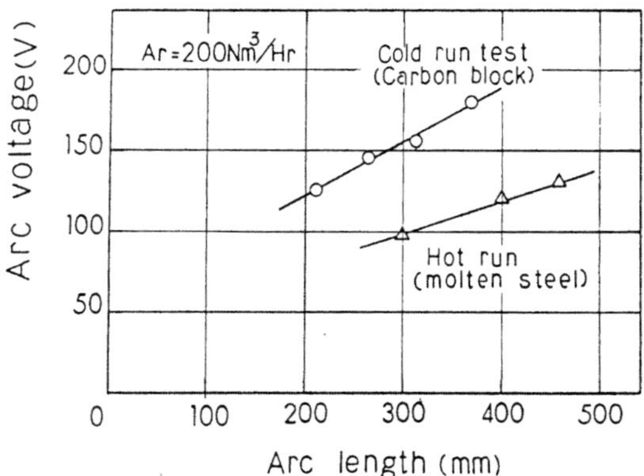

Figure 7.33: AC plasma arc length and voltage on molten steel and carbon block. [Ref. 13]

7.5.3.4 *Heating efficiency and temperature control*

Figure 7.35 shows the heat balance during plasma heating. About 65 % of the plasma heat was transferred to the molten metal, and about 22 % of the heat was lost to the cooling water. Thus, it is important to cover the torch surface with an insulating material to increase the heating efficiency. The plasma system is very effective in maintaining the superheat within a very close range. As shown in Fig. 7.36, the temperature variation, even in the non-steady state operation, was not more than 5 °C.

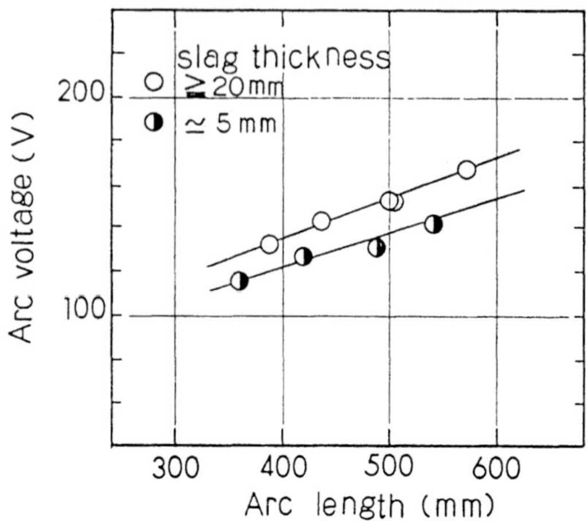

Figure 7.34: Effect of slag thickness on arc voltages and arc length. [Ref. 13]

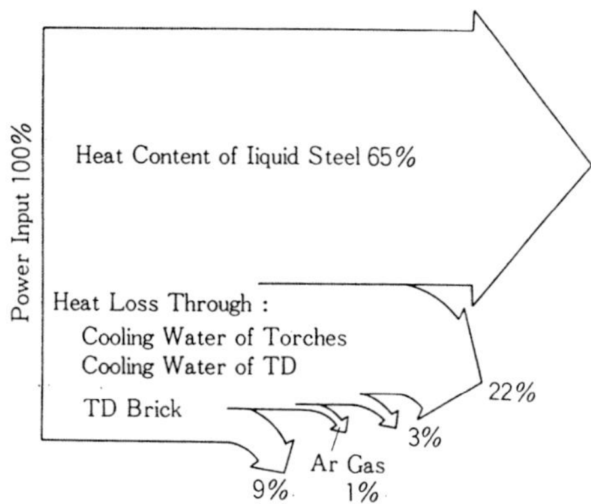

Figure 7.35: Heat balance during plasma heating. [Ref. 13]

The AC plasma heating implementation increased the cleanliness and improved the structure of the cast product. At the same time, the BOF tap temperature was reduced by 20 °C. An AC plasma system can be operated towards the end of a sequence in the non-transferred arc mode, which prevents any temperature drop. Higher temperature keeps the slag fluid, which makes slag pouring easier for a hot recycle tundish operation.

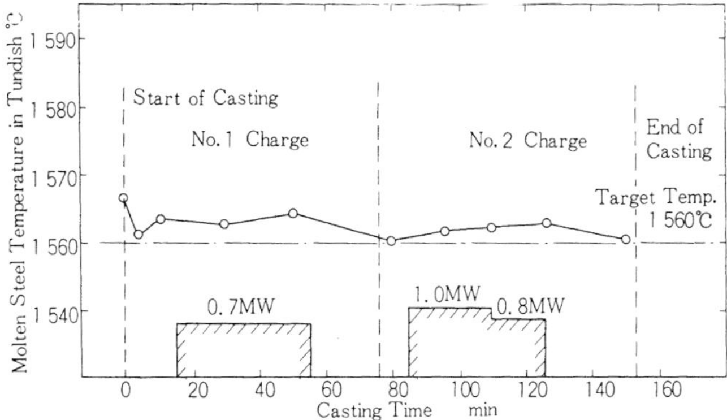

Figure 7.36: Steel temperature variation during plasma heating. [Ref. 13]

7.6 External Heating (Induction Systems)

Several steel companies [e.g. refs. 5, 8, 17, 18] have developed induction heating in tundishes of smaller capacity, normally smaller than 30 tons. These heating units are compact enough to be retrofitted to an existing tundish without any major modifications, and it may not even reduce the tundish capacity. A channel-type induction heater by Mabuchi *et al.* of Kawasaki Steel Corporation and ASEA (now ABB) is schematically illustrated in Fig. 7.37 [5]. It consists of an iron yoke or core, upon which a multiturn coil is wound. The other part is a refractory channel through which molten metal is forced to flow. This induction heater is a kind of transformer which is schematically shown in Fig. 7.38. The primary current is passed through the multiturn coil, which generates a magnetic flux in the core. This, in turn, induces a secondary current in the molten metal flowing through the channel. The induced current produces Joule's heat, which heats the molten metal. Since the molten metal is intensely stirred by electromagnetic force, it may cause severe erosion of the refractory in the induction channel. Thus, the channel refractory should be of a material with low wear rate and high thermal shock resistance. Suzuki *et al.* [8] at Nippon Steel Corporation tested properties of several refractory materials, and found that precast alumina with 10% SiC is a very suitable material for the channel. Sleeves of this refractory material were easy to install, low in cost, and had a high resistivity, which minimized heating by the induced current.

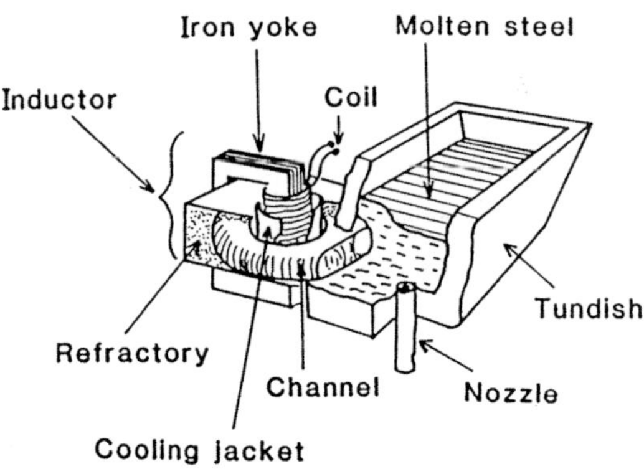

Figure 7.37: Channel type induction heating for tundishes. [Ref. 5]

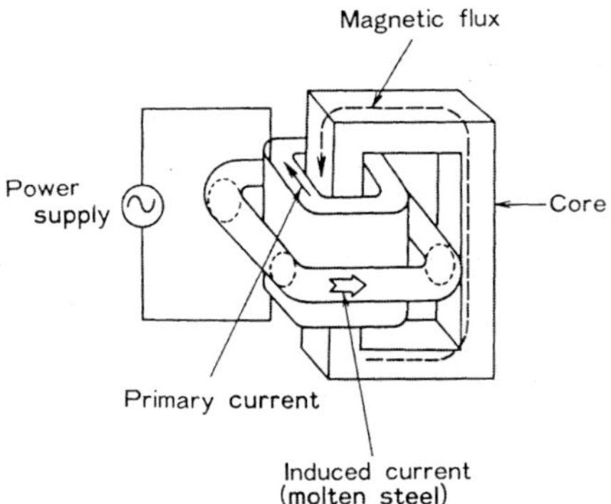

Figure 7.38: Transformer for induction heating in tundishes. [Ref. 8]

7.6.1 *Heating efficiency and temperature control*

The authors of Refs. 5, 8, 17, and 18 have reported that induction heating, generally, has a very high heating efficiency in the range of 80 to 90%. Suzuki *et al.* [8] used a 1000 kW induction heater in their L-shaped tundish, which is schematically shown in Fig. 7.39, and found its thermal efficiency to be 90%. About 5% of the heat was lost in the coil, core, cables, and power factor compensating capacitor. Another 5% was lost through leakage from the refractory sleeves. They measured the temperature of melt on the upstream and downstream of the heating system (plotted in Fig. 7.39), and found that the inductor heats the metal on both sides of the inductor. The efficiency on the upstream side was 20%, while it was 70% on the downstream side.

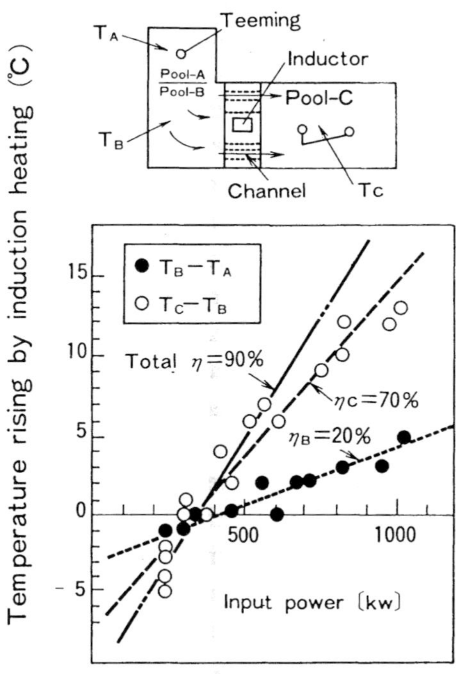

Figure 7.39: Temperature increase during induction heating as a function of input power. [Ref. 8]

Figure 7.40 shows measured temperature for a two-hour heat, which was controlled within ± 2.5 °C of the aim temperature by the induction heater. The dashed line is the predicted temperature drop of the melt if induction heating was not employed. Since the ladle stream temperature continues to decrease with time, power is gradually increased in order to maintain a nearly constant temperature. This has enabled the BOF shops [8] to achieve a 15 °C lower tap temperature at the BOF, and a 10 °C lower ladle temperature. Thus, the BOF and ladle refractory costs were reduced, and the metal yield was increased. With these benefits coupled with the improvements in bloom quality and structure, they confirmed that the installation of a heater in the tundish was fully cost effective.

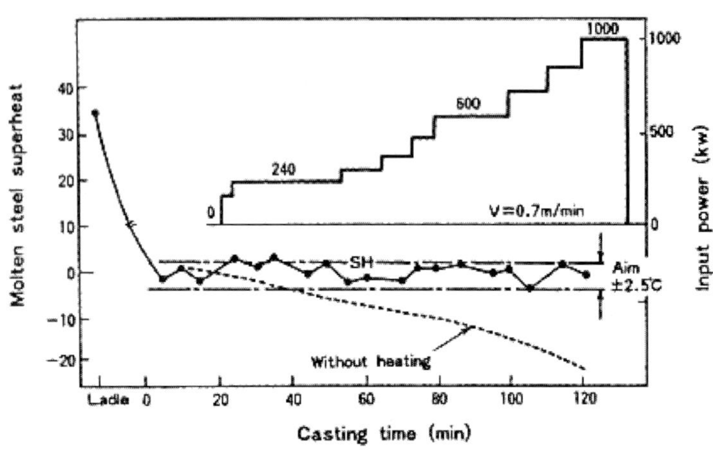

Figure 7.40: Temperature control during induction heating. [Ref. 8]

7.7 External Cooling

The steel melt temperature can be successfully lowered in the tundish by injection of steel powder. Hintikka and Jauhola [1] controlled the melt temperature in a tundish within a close range by steel powder injection at Raahe Steel Works in Finland. Their tundish was of 10-ton capacity with a melt depth of 1000 mm. A schematic of the operational set up is shown in Fig. 7.41. Steel powder was injected into the molten steel stream between the ladle and the tundish, which was enclosed in argon-filled ceramic shroud. The powder flow rate in their system could be varied from 5 to 35 kg/min with a carrier gas flow rate of 300 l/min. Being spherical in shape, the powder possessed excellent flow characteristics. Oxygen content of under 200 ppm, and carbon content of about 1 % may present some limitation in its use for certain grades.

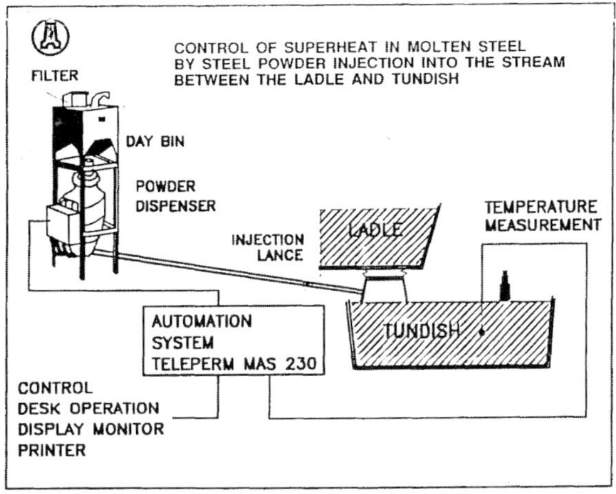

Figure 7.41: Set-up for steel powder injection for cooling melt. [Ref. 1]

If the powder could be injected into the metal stream, the cooling efficiency was found to be quite high (85 - 95%). A normal quantity of powder used varied from 0.5 to 1.0 wt %. Figure 7.42 shows that they could successfully achieve melt temperature in the range of ± 2 °C of aim temperature.

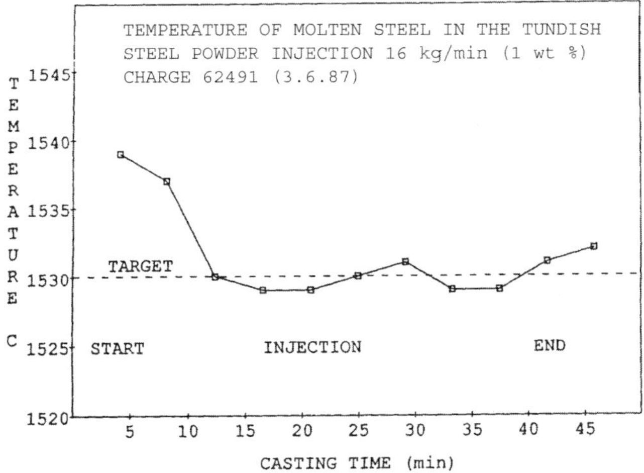

Figure 7.42: Melt temperature control during cooling by powder injection. [Ref. 1]

In spite of the stated success, powder injection is not a common practice in industry due to its limitations. The addition of metal powder to the melt causes local formation of solid steel crust which is not always easy to remelt and the cost associated with the process are the main limitations. More commonly, melt temperature is controlled in the ladle and brought to the desired temperature.

7.8 Concluding Remarks

Tundish melt temperature during casting of one ladle in a conventional process varies on an average of about 20 °C. The temperature drop is rather steep in the steady state casting periods during ladle changes. Control of melt temperature in a very close range is very beneficial for improving the quality and cast structure of the product. Additional benefits are that the BOF temperature could be lowered by about 20 °C, which reduces refractory cost in the BOF and ladle, and the metallic yield is increased. With continuous temperature measurement and plasma or induction heating, melt temperature can be maintained within a very close range. This technology is necessary for production of a very high quality of steel. Cooling of the melt can also be performed in the tundish, but it is not commonly done due to cost and quality concerns.

References

1. S. Hintikka and M. Jauhola: *The Sixth International Iron and Steel Congress, The Iron and Steel Institute of Japan Publication*, 1990, 3, 215-221.
2. T.J. Russo and R.M. Phillippi: *Steelmaking Conference Proceedings*, 1990, 73, 237-246.
3. S. Chakraborty and Y. Sahai: *Metall. Trans. B*, 1992, 23B, 152-167.
4. J. W. Miller and J. Hlinka: *Iron Steel Eng.*, 1970, 8, 123-133.
5. M. Mabuchi, T. Nozaki, Y. Habu, Yutaka Yoshii, B. Hanås, and J. Bostedt: *Steelmaking Conference Proceedings*, 1986, 69, 737-742.

6. H. Mori, M. Sawa, Y. Shia, M. Okabe, K. Kuwahara, and K. Sakai: *Tundish Metallurgy, Part II, ISS Publication*, 1991, 97-103.

7. K. Takeo and H. Iwata: Wire Journal, 1971, August, 32.

8. I. Suzuki, S. Noguchi, Y. Kashiwakura, T. Horie, and M. Saito: Steelmaking Conference Proceedings, 1988, 71, 125-131.

9. K. Matsumoto, Y. Hoshijima, K. Ishikura, K. Umezawa, Y. Nuri, and Y. Ohori: *The Sixth International Iron and Steel Congress, The Iron and Steel Institute of Japan Publication*, 1990, 3, 222-229.

10. H. Tanaka, H. Yamamoto, H. Kondo, J. Shoda, and Y. Suguro: NKK Technical Review, 1991, 62, 10-16.

11. N.J. Culp, Jr. and P.M. Cowx: *Electric Furnace Conference Proceedings*, 1989, 72, 107-110.

12. K. Inokuchi, A. Nagamune, and N. Ao: *Steelmaking Conference Proceedings*, 1993, 76, 245-249.

13. H. Iritani, T. Saito, H. Fujimoto, Y. Nakamura, M. Shimuzu, M. Kiyokawa, and S. Nishi: *The Sixth International Iron and Steel Congress, The Iron and Steel Institute of Japan Publication*, 1990, 3, 230-238.

14. H. Iritani, T. Saito, H. Fujimoto, Y. Nakamura, M. Shimuzu, M. Kiyokawa, H. Tokunaga, and H.J. Bebber: *Steelmaking Conference Proceedings*, 1991, 74, 699-705.

15. H. Iritani, J. Katsuta, T. Saito, H. Fujimoto, Y. Nakamura and N. Kiyokawa, *R & D Kobe Steel Engineering Reports*, 1991, 41, no. 4, 44-47.

16. T. Saito, K. Matsuo, H. Fujimoto, M. Maeda, K. Imiya and K. Umahoshi, *R & D Kobe Steel Engineering Reports*, 1990, 40, no. 1, 93-96.

17. A. Shiraishi, K. Iwata, H. Tomono, T. Nagahata, and A. Mori: *The Sixth International Iron and Steel Congress, The Iron and Steel Institute of Japan Publication*, 1990, 3, 264-270.

18. H. Nakata, J. Tsubokura, and H. Takahashi: *The Sixth International Iron and Steel Congress, The Iron and Steel Institute of Japan Publication*, 1990, 3, 470-477.

Chapter 8

Recent, Emerging, and Novel Technologies

8.1 Introduction

As detailed in the previous chapters, tundish technology has developed substantially to improve strand quality and caster productivity of premium steel at a decreased cost. Typical improvements made as of now include the following.

(1) Slag detectors to minimize ladle slag carry over into the tundish;

(2) A long nozzle or shrouding pipe with Ar injection to keep the inlet chamber inert for metal transfer from the ladle to the tundish, to avoid air reoxidation and slag entrainment;

(3) Large tundishes with sufficient depth and a reasonable height-to-width ratio;

(4) Control of melt flow and residence time by dams, weirs, baffles, and pads;

(5) Tundish heaters for consistent low superheat casting;

(6) Ar injection from the tundish nozzle, slide gate, and SEN to reduce nozzle clogging;

(7) Devices for quick automatic exchange of the tundish and SEN;

(8) Improvement of tundish operation with filled start, filler sand removal, inertization, tundish flux addition, and hot cycling; and

(9) Improvement of the tundish and nozzle refractory.

These improvements have survived the test of time, confirming their viability in terms of cost, productivity, and quality. They have come to near maturity in dealing with the steady state casting. In fact, a sealed tundish filled with inert atmosphere and of a reasonably large capacity of acceptable width to depth ratio with a bath depth of more than 1 m has proven capable of delivering clean melt to the mold without any additional contamination. The melt is poured from the ladle to tundish through an immersed long nozzle or a shrouding pipe. These measures, together with a hot tundish cycle with inert gas preheating and plasma or induction tundish heaters, have increased the productivity of continuous casters without much impairment of steel quality.

Problems still remain, however, for the transient periods of tundish operation including the opening of ladle, ladle change, ladle emptying, and grade changes. During the transient periods, most of the harmful macro inclusions of exogenous origins, *i.e.*, formed by reoxidation by air and entrainment of slag, generate in the tundish. These are due to the difficulties in maintaining an inert atmosphere for the melt stream from the ladle. Also, it is difficult to prevent emulsification of slag by the plunging melt stream in the tundish. Need for sequence casting of different grades in small quantities and for casting many heats in a sequence are continuously increasing. Thus, more advanced transition operation is required.

The occurrence of large alumina clusters, and accretion formation on the SEN, both are related to the indigenous fine alumina inclusion particles in Al-killed melt, particularly in Ti-bearing ULC grade, have not yet been solved satisfactorily. Nozzle clogging interrupts sequence casting and reduces caster productivity. It also causes detachment of the accretion and erratic asymmetric flow of the melt out of the SEN, entraining mold slag and inhibiting flotation of carried over inclusions in the mold. Turbulent flow, due to its random and fluctuating character, also leads to the formation of alumina clusters by agglomeration of fine indigenous alumina inclusions. These macro inclusions (and associated Ar bubbles, when Ar is injected into the SEN) are carried deep into the melt pool in the mold by the asymmetric flow, and get engulfed in the solidifying shell, impairing strand quality.

Although some solutions to these problems have been conceived and tested, they have not yet been fully accepted in industrial operation for several reasons. A single measure may not be sufficient for all outstanding problems. Instead, a combination of thoroughly tested and proven improvements in every sector of existing facilities and operations should accompany the development of new technologies. Many improvements were launched in the 1990's, but progress has been limited as of now. Some noteworthy examples are given in this chapter.

The productivity of a continuous casting machine (CCM) is generally far less than that of a blast furnace (BF), a basic oxygen furnace (BOF), and a hot rolling mill (HRM), even for a slab caster. If slab caster productivity is increased to a monthly average of 400,000 tons, an ideal combination of 1BF-1BOF-1LF-1CCM-1HRM would be most suitable for an integrated steel plant. The initial investment and operating costs of such an integrated plant would be kept to a minimum, and would contribute greatly to the competitiveness of the plant in the international market. In fact, recent developments have aimed at this goal [1]. Here, the tundish plays an important role in delivering deoxidized and clean melt to the CCM, heat after heat, without contamination, at designed temperature, and without any interruption or delay. To accomplish this, special care is needed for the hot and quick recycling of the tundish as a part of the integrated system. This issue will also be touched upon in the following.

8.2 Advances of the H-Shaped Tundish

The H-shaped tundish was a development from its predecessors, the LLTM (Ladle-Ladle-Tundish-Mold) and LLM (Ladle-Ladle-Mold) processes. The H-shaped tundish has been in operation since 1987 at the Nagoya Works of Nippon Steel Corporation and is not a new technology, but it is a unique one. The tundish has traditionally focused on preventing the clean ladle melt from forming exogenous macro inclusions of reoxidation- and slag emulsion-origin particularly during the transient periods of operation by:

(1) Separating the inlet and outlet compartments of the tundish by a refractory wall, and connecting the two compartments by a tunnel placed through the bottom of the wall; and

(2) Making overlapping teeming of the melt into the tundish possible from the preceding ladle and the succeeding ladle simultaneously, via a long nozzle without decreasing bath depth in the tundish during the ladle change.

The residence time of the melt in the tundish was made longer due to the extended length of the melt path. The preceding ladle could be emptied at a low rate near the end of the pouring. This would enable most of the melt to be poured into the tundish, and decrease carry over of ladle slag to the inlet compartment by vortexing and drainage. These characteristics served to minimize macro inclusions of ladle slag origin in the outlet compartment, and $[O]_{total}$ at the ladle change was kept constant, not increasing, as mentioned earlier in chapters 2 and 6.

The H-shaped tundish was then equipped with DC-arc plasma heaters as shown by Kimura *et al.* in Fig. 8.1 [1] to produce more demanding clean steels, such as deep drawn and ironing plastic film coated steel sheet. DC plasma heater made it possible to control the tundish melt temperature within ±5°C even during the ladle change, which in turn reduced clogging of the tundish and SEN.

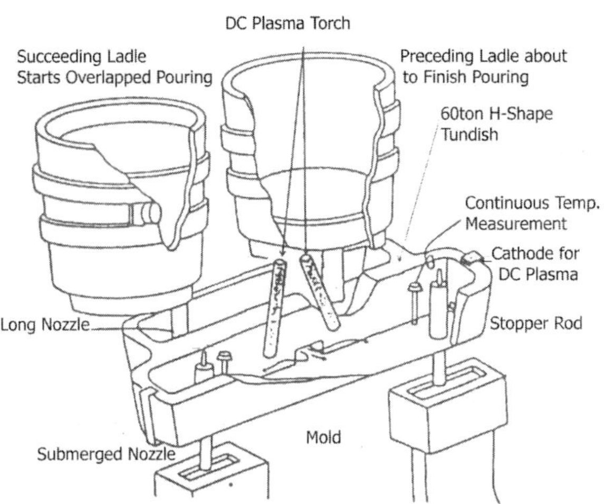

Figure 8.1: Schematic view of a 65-ton H-shaped tundish with two 2.35MW DC transfer Ar-N$_2$ plasma torches installed at the outlet compartments. [Ref. 1]

The defects of macro inclusions origin in the product coils were significantly decreased in the above-mentioned demanding class of steels, as shown in Fig. 8.2. It was also reported that industrial trials of Ar-bubbling in an H-shaped tundish were carried out, as shown by Takase *et al.* in Fig. 8.3 [2]. A porous plug for Ar gas injection was placed at the bottom of a tunnel which connected the inlet- and outlet-compartment of the tundish. This is the place where melt flow rate is the fastest in the tundish. Bubble diameters were found to be acceptably fine, as shown in Fig. 8.3(b), and defects arising from inclusions were considerably decreased, as shown in Fig. 8.3(c) Grade changes were also made with extremely low melt level practice in the tundish at the end of the preceding heat by utilizing good separation of the two heats with the two compartments and tunnel configuration.

Presumably, the high refractory cost is a drawback to making the H-shaped tundish popular for non-demanding grades. However, if this tundish is modified to accommodate the hot tundish cycle operation, the drawback would be minimized.

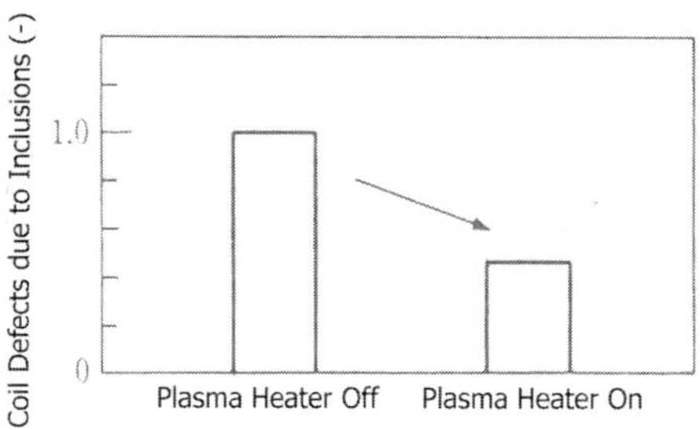

Figure 8.2: Effect of plasma torch heating of the melt on the defects of macro inclusion origin in cold rolled coils. [Ref. 1]

8.3 The Centrifugal Flow Tundish (CF Tundish)

An interesting application of centrifugal force acting differently on the steel melt and inclusions upon rotation of the melt, resulting in inclusion separation to the bottom surface of rotating concave shape melt, was commercialized to cast stainless steels in a one-strand slab caster. Melt rotation was implemented in production at Chiba Works of Kawasaki Steel Corporation (now JFE Steel) first with a 10-ton CF Tundish to confirm the results obtained earlier with an experimental 600 kg CF Tundish. It was later scaled up to a 30-ton CF-tundish that consisted of a 7-ton rotation chamber (inlet compartment) and a 23-ton rectangular chamber (outlet compartment) as shown by Miki *et al.* in Fig. 8.4 [3]. Aluminum killed stainless steel melt or high carbon steel melt (S45C) was poured from a 180-ton ladle into the rotation chamber where it was rotated by a progressive magnetic field generated by a semi-cylindrical linear motor type electromagnetic stirrer attached outside the compartment. The two compartments were connected by a through hole placed at the bottom.

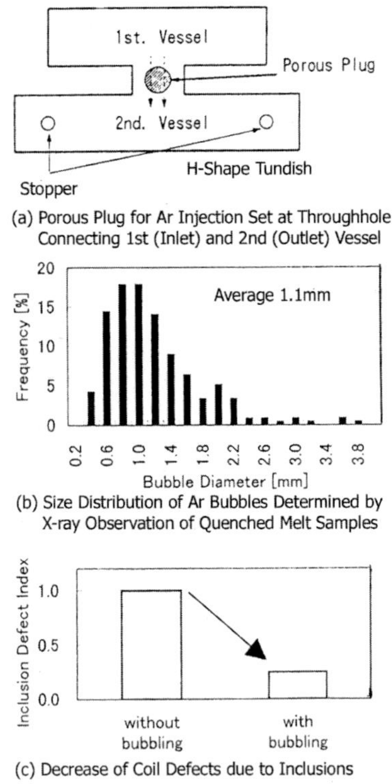

(a) Porous Plug for Ar Injection Set at Throughhole Connecting 1st (Inlet) and 2nd (Outlet) Vessel

(b) Size Distribution of Ar Bubbles Determined by X-ray Observation of Quenched Melt Samples

(c) Decrease of Coil Defects due to Inclusions

Figure 8.3: Effect of fine Ar bubble injection on the defects in cold rolled coils. [Ref. 2]

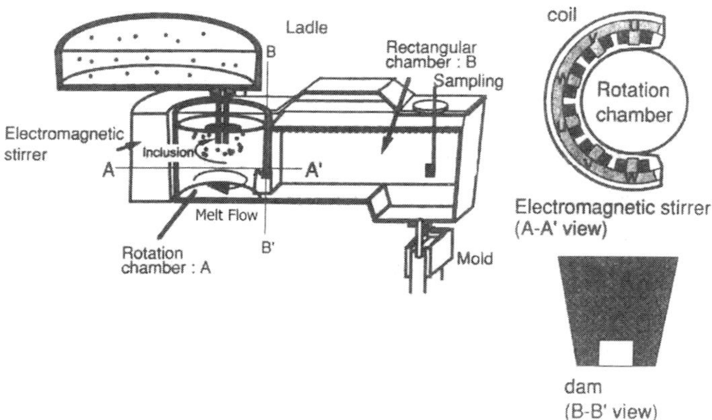

Figure 8.4: Schematic view of a 30-ton centrifugal flow tundish (CF Tundish) with a semi-cylindrical progressive magnetic field stirrer on a 7-ton inlet compartment (rotation chamber). [Ref. 3]

Production data indicated that macro inclusions during the ladle change of Al-killed and VOD treated ferritic stainless steel melts (SUS430) decreased almost by 50% by rotating the melt at 45 rpm. This is shown in Figs. 8.5 [3] where the peak value of macro inclusions due to slag entrainment during ladle change (see curves for 12- and 29-ton after the ladle change) reduced to much smaller values. Also, $[O]_t$ in steady state casting decreased from about 15 ppm to 7-8 ppm by the rotation. As dissolved oxygen content was about 3 ppm in this case, finer sized inclusions decreased to about one half. This corresponded to the noticeable decrease of inclusions smaller than 4 μm, as shown in Fig. 8.6 [3]. Surface defects of resulting hot- and cold-rolled coils decreased to 1/2 and 1/3, respectively, as shown in Fig. 8.7 [3]. Similar results were also obtained for S45C. The apparent volumetric deoxidation rate constant, K_o, for the rotated melt in the inlet compartment was found to fall at the high end of the reported correlation between K_o and stirring energy density, ε, shown in Fig. 8.8 [3]. Here K_o and ε are given by the following equations:

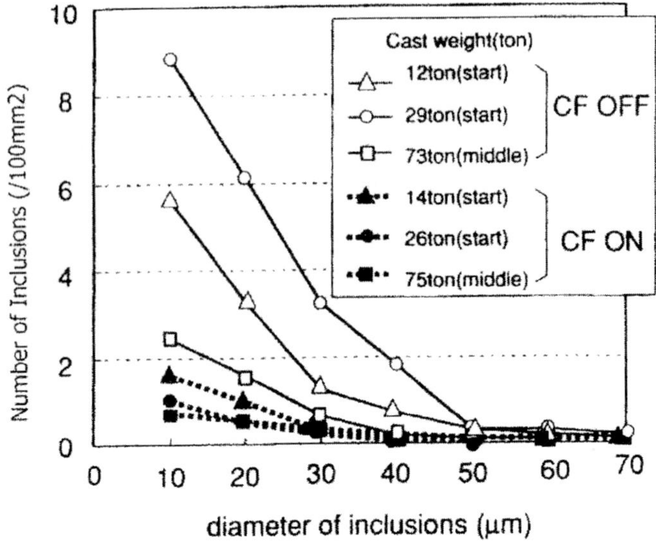

Figure 8.5: Effect of centrifugal rotation on the macro inclusion content in a CF tundish in sequence casting of Al-killed ferritic stainless steel. [Ref. 3]

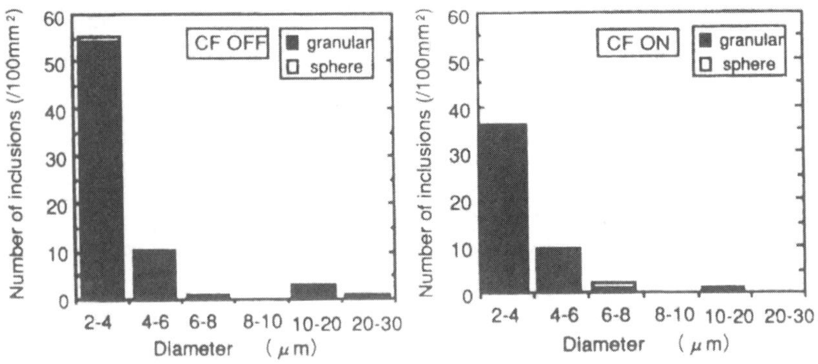

Figure 8.6: Effect of centrifugal rotation on the macro inclusion content of Al-killed ferritic stainless steel. [Ref. 3]

$$\ln d[O]_t / dt = -K_o ([O]_{t=t} - [O]_{t=\infty}) \tag{8.1}$$

$$\varepsilon = 2\pi NT/W \tag{8.2}$$

where N is the rotation rate, T is torque, and W is the weight of the melt. This K_o of 0.7 was much greater than 0.04 for the rectangular chamber of conventional tundish. Such an effective removal of macro- and micro-inclusions was attributed to the coarsening/agglomeration of the inclusions by the centrifugal force.

As size of the semi-cylindrical magnet was not excessively large and its effect on exogenous- and indigenous-inclusion removal was considerable, the CF Tundish is well suited for casting high quality specialty steels. In particular, it is very good for casting Al-killed steels since Al_2O_3 inclusions agglomerate better due to their higher interfacial tension with liquid steel. Cost remains a problem for its acceptance for Si-Mn deoxidized grades, and the difficulty to use a tundish hot cycle operation also can be a limitation of the process.

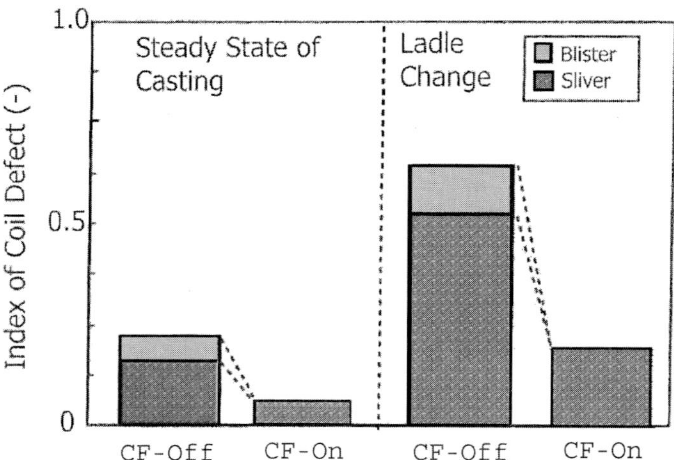

Figure 8.7: Reduction of the defects on cold rolled coil by centrifugal rotation in a CF tundish. [Ref. 3]

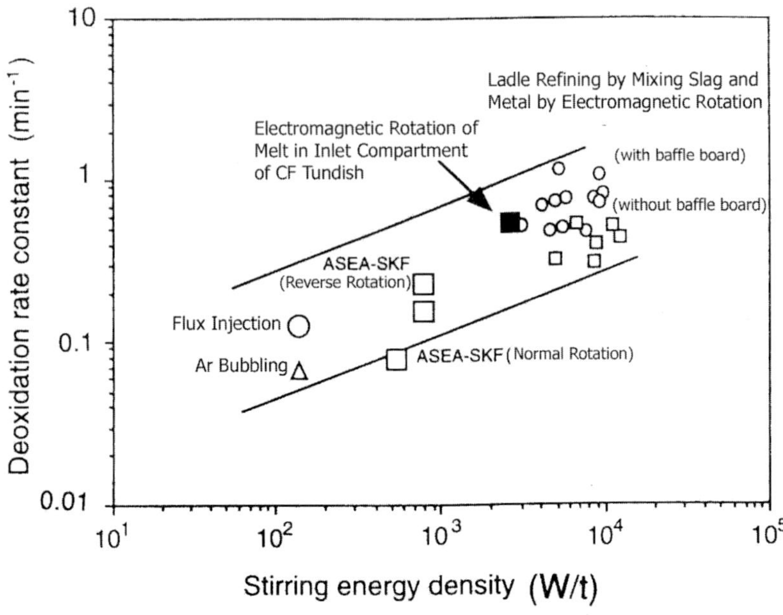

Figure 8.8: Volumetric deoxidation rate constant in the inlet compartment of a 30-ton CF tundish as a function of the dissipation rate of stirring energy. [Ref. 3]

8.4 Argon Bubbling in a Tundish

The inert gas/melt interface that is provided by gas bubbles injected into the steel melt has proved to be a favorable site for Al_2O_3 inclusions to adhere. The adhered inclusions instantly agglomerate into large clusters as revealed in-situ by Yin *et al.* [4]. Stirring of the melt by inert gas bubbles also enhances collision and agglomeration of inclusions suspended in the melt. As shown by Okamura *et al.* [5], the stirring also helps in transferring the inclusions to the bubble/melt interfaces and subsequent flotation of the bubbles with the adhered inclusions to the surface of the melt. On the contrary, the Ar injection rate in a tundish needs to be limited to avoid emulsification of tundish slag and re-entrainment of surfaced inclusions near the slag/melt interface. It is, therefore, necessary to generate fine Ar bubbles to create the largest possible bubble/melt interface for inclusion removal with the limited rate of injection. Thus, injecting fine Ar bubbles into the melt in a tundish under non-oxidizing conditions is an effective way to reduce the inclusions, both exogenous and fine indigenous types, making the tundish a good refiner. Early trials were to inject Ar bubbles from the bottom of the inlet or outlet sections in a closed tundish to make the atmosphere inert, as discussed in Chapter 6. Injection of Ar bubbles is particularly beneficial when the melt flow rate through the tundish is increased with increasing casting speed, at which point flotation of the inclusions becomes difficult.

Many attempts were made to realize such Ar bubbling in a tundish, but seemingly have not become popular. This is due to the difficulty inherent in keeping homogenous dispersion of fine Ar bubbles consistent throughout the long sequential casting at an acceptable cost. The contact angle between the steel melt and refractory at the three-phase boundary, Ar gas/steel melt/refractory, at the Ar gas inlet is large, and hence the minimum size of the bubbles available under normal gas injection conditions remains larger than desired. Also, injected Ar bubbles, even when they are made fine, tend to coalesce together to make large bubbles immediately above the point of injection.

The size of Ar bubbles injected into the melt flow has been known to decrease with the thickness of velocity boundary layer of melt flow at the point of the injection. The boundary layer gets thinner as the flow velocity increases. Injected Ar bubbles can grow only within the boundary layer to a size proportional to the boundary layer thickness. When the bubbles get larger than that, they are forced to detach by the melt flow. Thus, fine bubbles can be obtained by increasing the melt flow velocity at the point of Ar gas injection. This phenomenon was used in a tundish by injecting Ar from a rotating plug immersed in the melt as cited in Chapter 6. Unfortunately, refractory cost and operational difficulties made this method unpopular despite its success in reducing inclusions and SEN clogging. Use of a porous plug with very fine pores was another solution that was attempted with some success as cited before. The drawback in this case involves the difficulties in hot cycling of the tundish.

If small-sized Ar bubbling in a tundish becomes successful in the long run, it will be a promising way to produce ultra-clean steels. To make Ar bubbling an industrially viable technology, the durability of materials for Ar gas injection needs to be improved for long lasting sequence casting. Also, the structure of the injection unit should be simple and easy to exchange for a tundish hot cycle. As mentioned in section 8.2, for an H-shaped tundish, Ar bubbling was utilized for enhancing inclusion removal at the through hole. In plasma heating, it was also used for improving heat transfer in the plasma heating chamber, where no tundish slag exists and the atmosphere is totally inert. In these cases, slag emulsification and open eye formation do not occur and are not problems. Additional improvements of the refractory structure to form fine argon bubbles on the gas injector are required to minimize the risk of tundish flux entrainment.

8.5 Electromagnetic Control of Melt Flow

Application of an electromagnetic field to brake or accelerate melt flow in a continuous casting mold has been quite successful in reducing inclusions and improving surface defects of strands. Electromagnetic braking reduces the penetration depth of the melt flow from the SEN into the melt pool in strand, suppresses meniscus turbulence, prevents entrainment of mold slag, reduces the occurrence of asymmetric flows from the SEN, and maintains a proper flow velocity and temperature near the meniscus. These effects have contributed greatly in decreasing the occurrence of inclusions and surface defects, particularly at high casting speeds. Various kinds of electromagnetic flow controllers for such applications have found increasing popularity in highly productive modern continuous casters. The controller systems are called the EMBr (electromagnetic braking) and FC mold (flow control mold), which employ a static magnetic field; and the EMLS (electromagnetic level stirrer), which employs a progressive magnetic field to counteract the melt flow. All these controllers decelerate the melt flow.

An alternative type of electromagnetic flow controllers has also been in industrial use, which accelerates the melt flow in front of the solidifying shell to wash inclusions and enriched solutes away from the growing shell and produces the seeds for equiaxed crystals by fragmenting dendrite arms. The EMLA (electromagnetic level accelerator) and MEMS (in-mold electromagnetic stirrer) for slab casters and many of their variants with progressive and rotating magnetic field for bloom/billet casters fall into this category. They are intended to reduce subsurface inclusions and to reduce the center segregation of solutes. Further extension of these technologies has been directed towards the tundish and SEN in recent years, with the following preliminary results.

In the tundish area, an early attempt was made actively to control melt flow by applying a static magnetic field horizontally to a model tundish containing mercury or Wood's metal. The experiments were aimed at improving the melt flow in an open channel in the tundish and retarding the onset of vortexing of the melt at the tundish nozzle during ladle change. Both these flow improvements were intended to enhance the flotation of macro inclusions and to reduce the entrainment of tundish slag into mold. It was found that the velocity distribution became uniform, as shown by Takeuchi *et al.* in Fig. 8.9 [6], and consequently the critical depth of vortexing (onset of drain sink) was reduced at a field of 0.1 T. The improvements were further confirmed in a hot model experiment with a 900 kg steel melt, as shown by Idogawa *et al.* in Fig. 8.10 [7]. The figure indicates that, if a tundish were modeled to consist of a perfect mixing region (V_M) and plug flow region (V_P) connected in series, application of a static magnetic field (0.15 T) vertical to the flow increased the volume fraction of the plug flow region from 0.025 to 0.594, which improved the flow characteristics.

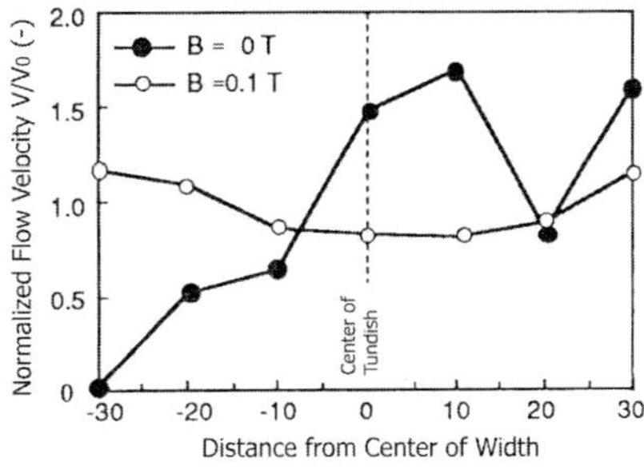

Figure 8.9: Normalized flow velocity of mercury in a model tundish by applying a horizontal static magnetic field. [Ref. 6]

 Tundish Technology for Clean Steel Production

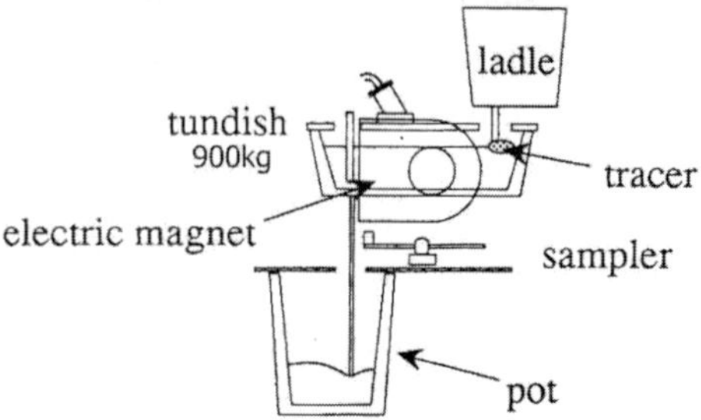

Schematic view of the equipment

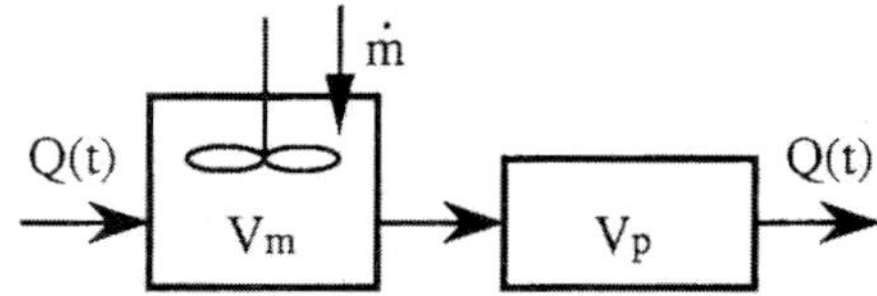

Perfect mixing region   Plug flow region

Representation of volume fractions
determined by the mixed flow model

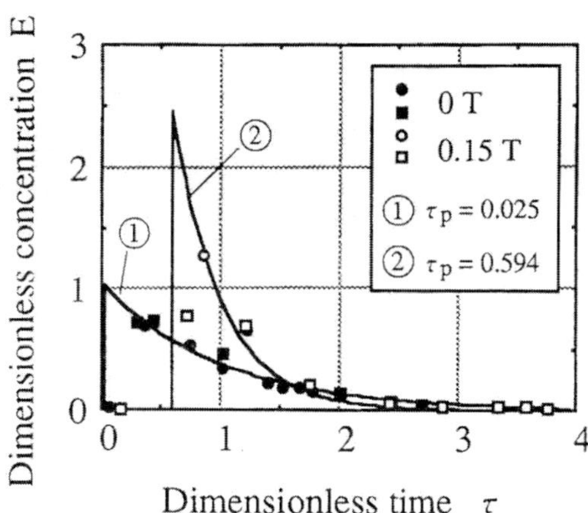

Effect of magnetic field on the residence
time distribution of liquid steel

Figure 8.10: Increased plug flow volume in a 600 kg tundish by applying a static magnetic field. [Ref. 7]

The effect of the magnetic field on vortexing at the tundish nozzle during the ladle change at low melt depth in a tundish is shown in Fig. 8.11 [6]. Application of a magnetic field either vertically (circles) or horizontally (lines) could retard the onset of the vortexing beyond the magnetic field of 0.1T. A horizontal magnetic field (lines), however, produced a more consistent effect than the vertical field (circles).

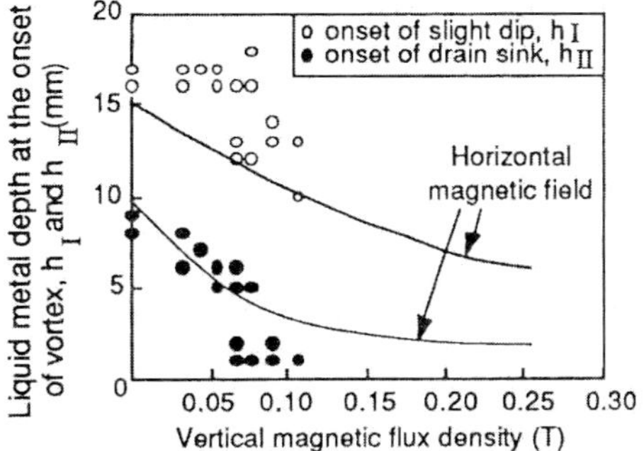

Effect of magnetic flux direction on the suppression of dip and vortex of molten metal.

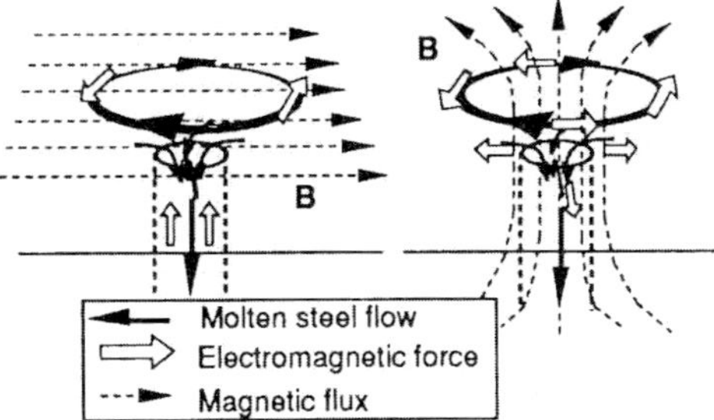

Figure 8.11: Suppressing the onset of vortex at the tundish nozzle by applying a horizontal and vertical static magnetic field to mercury in a model tundish. [Ref. 6]

The melt flow out of a tundish can also be controlled by a magnetic field. A calculation indicated that an EMBr type static magnetic field of 0.5T applied to a SEN over a length of 300 mm between the slide gate and the meniscus of melt in the mold prevented the occurrence of asymmetric flow out of a bifurcated SEN, as shown by Morishita *et al.* in Fig. 8.12 (see curve for B1 = B2) [8]. It was also possible to control

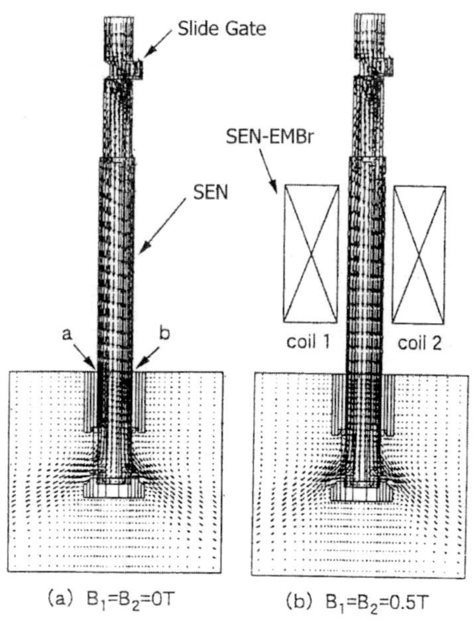

Velocity distribution in submerged nozzle with and without EMBr

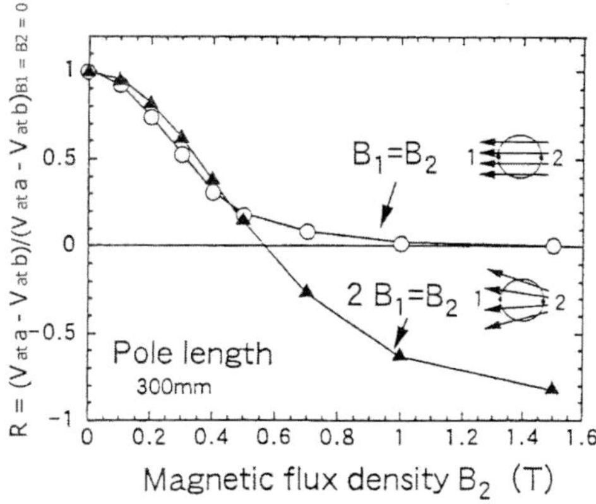

Figure 8.12: Simulation of controlling biased flow exiting from a bifurcated SEN by applying a static magnetic field to the SEN. [Ref. 8]

the degree of the melt flow asymmetry by applying a static magnetic field gradient at B2 of 1.0T as shown in the same figure (see curve for 2B1 = B2). Space limitation and the difficulties in constructing a system to monitor the asymmetry of the outlet flow from a bifurcated SEN must be overcome before the idea to feed back the asymmetry information and activate the magnetic field gradient to prevent the asymmetry can be materialized. Competition of the magnetic field application with other methods to minimize nozzle clogging is a separate issue.

Another indirect application of the magnetic field to the SEN was to replace a bifurcated- or multihole-SEN with a straight SEN to avoid the deposition of Al_2O_3 inclusions near the exit ports, and to utilize the electromagnetic braking to limit the penetration depth of the melt. To counteract deep penetrating flow out of the straight SEN, a static magnetic field of 0.30 T was applied to the straight SEN in a production set-up in two stages by FC Mold as shown by Nara *et al.* in Fig. 8.13 [9]. Observed and calculated results were obtained for a feed rate of 3.5 ton/min, a mold size of 260 mm x 1080 mm, and Ar gas injection of 7 Nl/min into the tundish nozzle. In the mold, the nozzle exhibited satisfactory decrease of flow velocity. The calculated flow pattern is shown in Fig. 8.13, indicating considerable braking of the downward penetrating flow. Sequential casting of three 230-ton heats of ultra-low carbon, Al-killed steel showed no deposition of Al_2O_3 inclusions in the straight SEN and no slab surface defects. In particular, blow holes arising from the Ar injection from the tundish nozzle to reduce nozzle clogging were much improved, as shown in Fig. 8.14 [9]. Thus, a magnetic field can be a potential means of avoiding nozzle clogging and improving cast metal quality. Application of an electromagnetic field is expected to provide the tundish and SEN with great potential for active control of the melt flow and strand quality. Cost issues and limited space availability for the installation of the device may impose some obstacles.

8.6 Tundish Heaters

Tundish heaters have been developed, as mentioned in Chapters 6 and 7, to compensate for the temperature drop of steel melt at the start, junctions, and at the end of heats in a sequential casting. The tundish heater minimizes variation of temperature over the entire heat sequence, and serves to reduce the tapping temperature of the BOF or LF. It also makes consistent low superheat sequential casting possible, which increases equiaxed crystal fraction in the pool end in strands. Increased equiaxed crystal fraction is favorable to reduce ridging and roping in ferritic stainless- and silicon-steel sheet. It also reduces the harmful center segregation of solutes in slabs for HSLA and high carbon steel plates or coils, and in blooms or billets for high carbon steel tubes, rods, and wires, when accompanied with the soft reduction of the pool end in the strand.

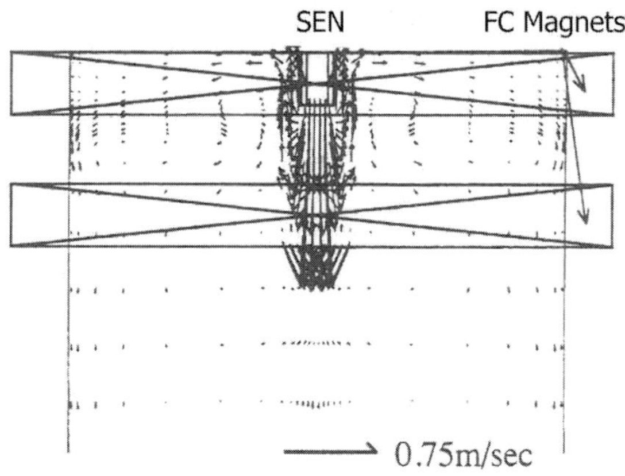

Figure 8.13: Damping melt flow through the SEN with a straight single spout by applying a two-stage static magnetic field in an FC mold. [Ref. 9]

Tundish heaters are, as mentioned earlier, largely classified into two types, coreless induction channel heaters and plasma heaters. The former type is preferred for a small size tundish with lower melt throughput rates for a long casting period. The latter type utilizes either AC- or DC-arc plasma in a single- or twin-torch installation. For induction heating, the increase of refractory consumption due to wear by aggressive flow in the channel is an important issue. The cost and labor of relining the complicated channel structure needed in induction heating should be minimized. On the contrary, electromagnetic pinching and turbulence of the melt in the channel are shown to enhance collision, agglomeration, and removal of inclusions. Also, the heating efficiency of about 90% for the induction heaters is better than that of plasma heaters. Induction heaters have long been in operation in many plants, and have become more or less a mature technology. Thus, there has not been much revolutionary progress reported recently.

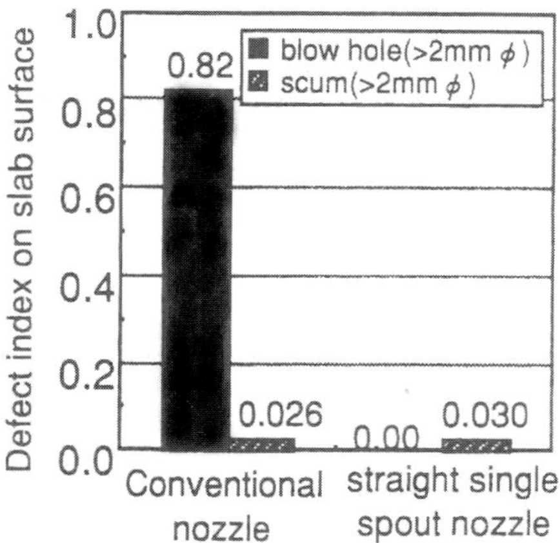

Figure 8.14: Defects on slab cast in FC mold (260mm x 1080mm) at 0.3T at 3.5ton/min via a straight single spout with an Ar injection of 7Nl/min. [Ref. 9]

For plasma heating, care was taken to maximize the energy input efficiency from the plasma to bulk of the melt by controlling the melt flow in the tundish in order to transfer the heated melt from the surface into the bulk with tundish furniture (usually a dam) and/or Ar-bubbling. Also, nitrogen pickup into the melt needs to be suppressed by preventing any aspiratory intake of ambient air by Ar-plasma into the heating compartment. The DC-transfer plasma type is less noisy than its AC-counterpart. Single DC-transfer torch installation is common for small size tundishes, with an anode plate embedded in the tundish refractory for space and cost limitations. Twin DC-transfer torch installation is favored for high throughput, large tundishes for high productivity casters. Twin torch installation delivers higher energy input, and does not require the embedded anode plate in the tundish lining or the connecting wiring to the anode plate, which are inconvenient for the hot cycling of a tundish. Parallel delivery of electric current to the twin torches reduces the otherwise strong magnetic field, decreasing noise to the peripheral instrumentation. A comparison of recent single- and twin-installations is listed by Kittaka *et al.* in Table 8.1 [10]. By the use of two 1.1 MW twin DC-transfer type torches that were set in a 65-ton tundish, as shown in Fig. 8.15, (also see Ref. [11] by Yamada *et al.*) melt temperature fluctuation at the ladle change was decreased to ±5°C, as shown in Fig. 8.16 [10]. The heating was effective in reducing the nozzle clogging of the SEN as shown in Fig. 8.17 [10].

Table 8.1: Comparison of a single- and twin-DC transfer plasma torch for tundish heating. [Ref. 10]

	Plasma I		Plasma II	
	Single torch		Twin torch	
Plasma torch type	Cathode torch (Anode: TD)		Cathode and Anode torch	
	DC			
Composition	Ar gas / Cathode / Nozzle / Heated object (Anode)		Ar gas / Anode / Cathode / Nozzle / Heated object	
Maintainability of torch	○	An external diameter: 1.1 (Type of separated nozzle)	○	An external diameter: 1.0
Power	○	0.2~0.7MW Recommend 0.5MW	◎	0.3~0.8MW
Heating efficiency	○	60~70%	○	60~70%
Plasma gas → noise	◎	1.0 noise: Low	○	2.4 noise: a little low
Lifetime of torch	○	Cathode: 1.0 Anode: 1.0	◎	Cathode: 2.0 Anode: 2.0 Nozzle: 2.0
Electric noise	△	Generated noise: a little	○	Generated noise: little
Revamp	△	Heating chamber Anode plate in tundish	○	Heating chamber
Maintainability of refractory in tundish	△	Need to maintain anode plate in tundish Adapted TD heat recycle: difficult	○	Only general maintenance Adapted TD heat recycle: possible
Equipment layout	◎	Small (Because of one torch)	○	Middle (Because of two torches)

Today, the channel inductor type heaters are used largely by casters that produce blooms for high quality long products, whereas twin torch type plasma heaters are equipped with slab casters that produce slabs for demanding products and/or narrow slabs that require long casting time. In both cases, the increase in caster productivity and premium product yield for transient blooms and slabs, and the decrease in BOF tapping temperature rationalize the investment and running cost of the heaters.

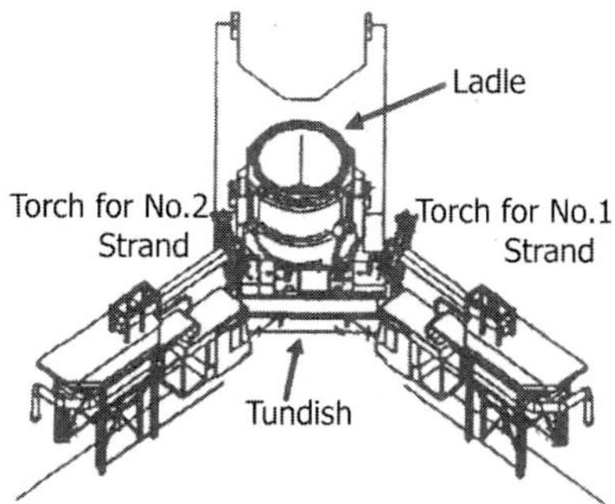

Figure 8.15: Two 1.1MW DC transfer plasma torches on a tundish for a two-strand slab caster. [Ref. 10]

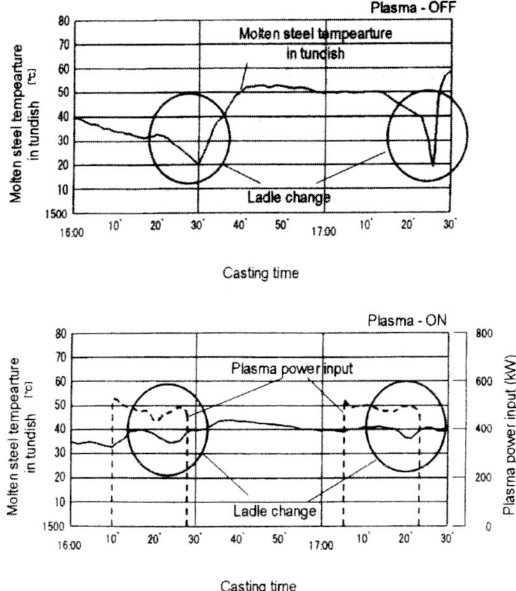

Figure 8.16: An example of improved melt temperature fluctuation at ladle changes by plasma torch heating. [Ref. 10]

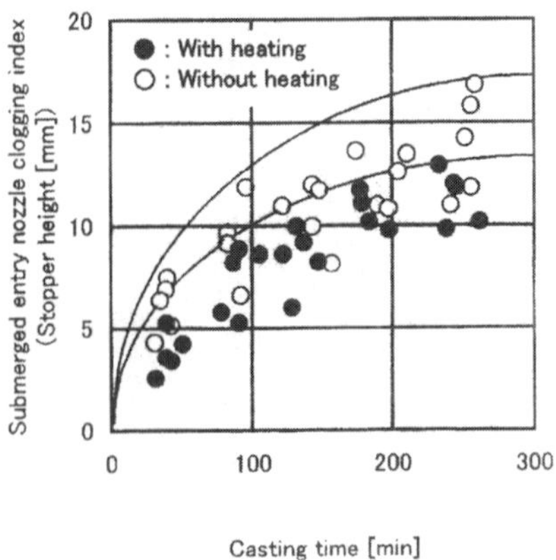

Figure 8.17: Effect of plasma torch heating of melt in a tundish on the reduction of SEN clogging, as measured by stopper head displacement during casting. [Ref. 10]

8.7 Hot Cycling with a Single Tundish Reheated Under Inert Atmosphere

As mentioned in Chapter 6, hot cycling of a tundish was effective in decreasing (1) the temperature drop of steel melt during transient periods of casting; (2) tundish refractory consumption; and (3) the labor for relining the tundish. However, hot recycling suffered from the steel melt contamination by iron oxides arising from the reoxidation of solid steel retained in steel/slag mixture that adhered to the tundish during long sequential casting. The remaining solid steel in the steel/slag mixture in a hot tundish gets reoxidized during reheating for the subsequent use. The iron oxide drips down onto the tundish bottom and reoxidizes the incoming melt during the early half of the next heat, causing the formation of macro inclusions.

To avoid this, attempts were made to melt away the adhered mixture during casting by adding a flux with low melting temperature to the tundish before reheating. Even then, the removal of the mixture was found to be inadequate. Accordingly, reheating of the prepared hot tundish was made under a non-oxidizing atmosphere, first with preheated nitrogen gas and later with preheated reducing hydrogen-nitrogen gas mixture at Kawasaki Steel Corporation (now JFE Steel) as shown by Hara *et al.* in Fig. 8.18 [12]. The inert atmosphere and reducing reheating conditions effectively reduced the reoxidation and the resulting sliver defects of the first heat in the subsequent casting sequence [13].

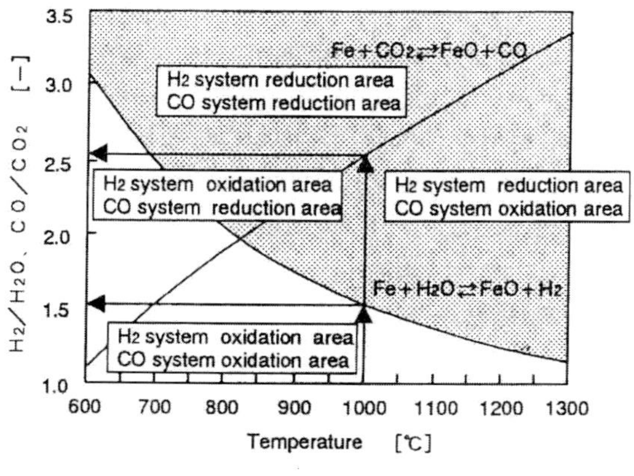

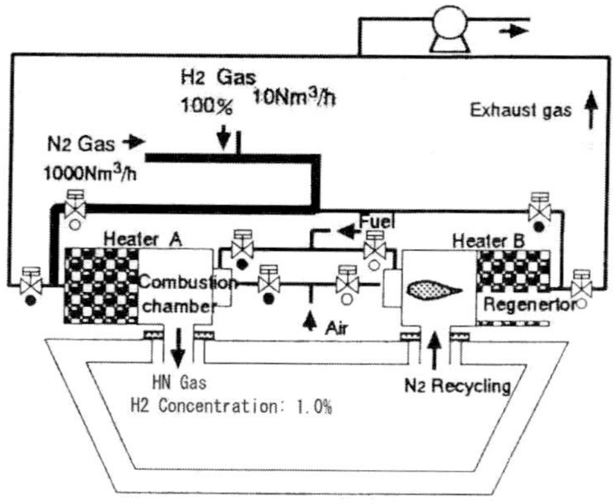

Figure 8.18: Preheating a tundish by injecting reducing H_2-N_2 mixed gas preheated in a recuperator chamber. [Ref. 12]

In conventional operation, two tundishes were used alternatively, one was employed for casting the preceding sequence and the other was reheated and prepared for the subsequent sequence, as shown in Fig. 8.19 (a) [14]. Reheating time was generally long, which increased the reoxidation of the steel/slag mixture. However, further improvements in the design of the tundish nozzle and equipment have enabled the replacement of tundish nozzle in less than 20 min. This made it possible to recycle the hot tundish for multiple sequential castings with only one tundish, as shown by Kooriyama *et al.*[14] in Fig. 8.19 (b). Service life of the tundish with the hot recycle operation was made 3.5 times longer than that for the conventional tundish operation.

The single-tundish hot cycle operation coupled with H_2-N_2 gas reheating greatly reduced the heat loss, reheating time, and the reoxidation of steel in the mixture during the tundish preparation. As a consequence, exogenous macro inclusions of reoxidation origin, and hence rejects due to sliver defects on coils cold rolled from the first heat, were significantly decreased as shown in Fig. 8.20 [14]. An additional advantage of the hot cycle tundish is that the temperature drop at the heat junctions could be minimized owing to high sensible heat retained in the tundish refractory, sometimes making it unnecessary to implement tundish heaters.

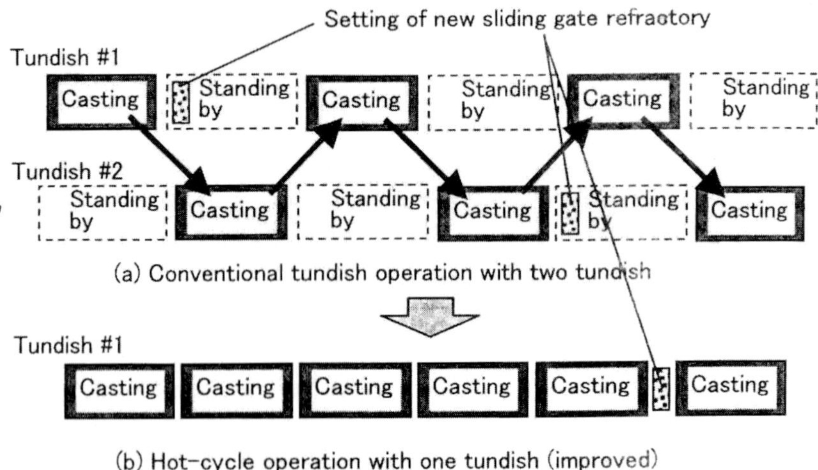

Figure 8.19: Single-tundish operation compared with conventional two-tundish operation for sequential casting. [Ref. 14]

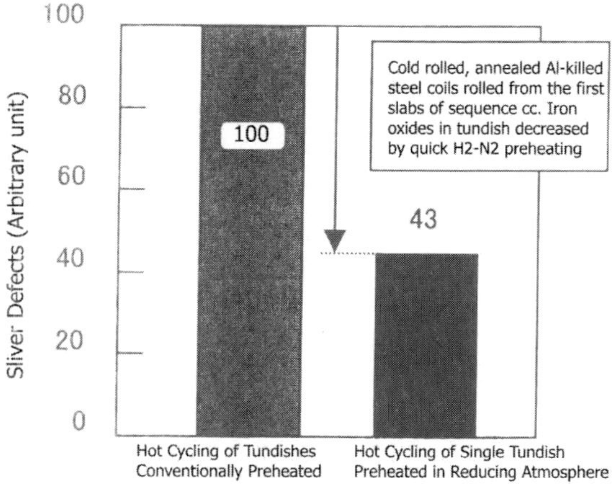

Figure 8.20: Decrease of surface defects on coils of macro inclusion origin by use of a single tundish for hot cycling. [Ref. 14]

8.8 Improved Argon Shrouding of the Melt Stream from the Ladle to the Tundish

The long nozzle and the shrouding pipe (including their variants) have been the two common means to protect steel melt from reoxidation by air during melt transfer from the ladle to the tundish. Each of these methods has its advantages and disadvantages. Long nozzles have found more popular use because of their convenient automatic mounting, their small space occupancy, and their isolation from the tundish structure. However, the long nozzle needs to be replaced heat after heat. In case the ladle opening fails, the long nozzle needs to be removed for oxygen cleaning of the ladle nozzle. The submerged opening of the nozzle requires care to minimize aggressive blowbacks. Ladle opening with a long nozzle under an inert atmosphere for the first heat may be more difficult.

A shrouding pipe (alternatively named a pouring tube, etc.), shown in Fig. 8.21 [15], does not require replacement and removal after each ladle, and shrouding with Ar for the first heat is easier. However, Ar gas consumption, accretion of melt splash on the inner surface of the pipe, and entrainment of slag floating on the meniscus in the shrouding pipe by incoming melt stream have been major problems. Nevertheless, recent improvements at Sumitomo Metal Industry, shown by Kasai *et al.* in Fig. 8.21c as *'improved'* [15], changed the location of Ar injection from the top inlet port (Fig. 8.21a) or side inlet port (Fig. 8.21b) to the semi-immersed pores near the meniscus in the shrouding pipe.

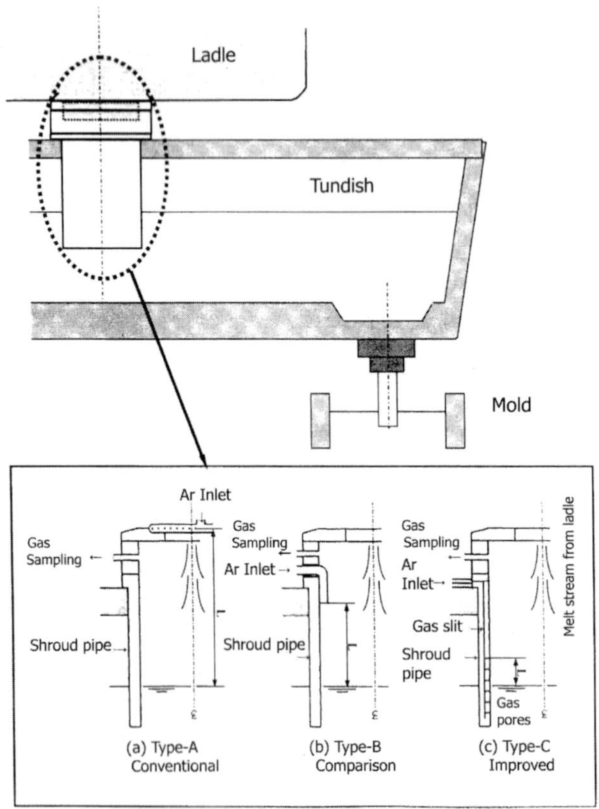

Figure 8.21: Conventional shrouding pipe with Ar gas injection and its improved version for quality benefits. [Ref. 15]

Aggressive mixing of the downward flow of Ar with the remaining air in the shrouding pipe for the top or side inlet set-up was reduced in the *'improved set-up'*, which pushed the air out to attain an inert atmosphere more quickly during the transient operation. This effect is clearly observable in Fig. 8.22 [15], where 600-1200 Nl/min of Ar was injected from above and below the meniscus in the pouring tube of the inlet chamber for a 32-ton tundish. About 40 or 50% reduction of macro inclusions (>120μm dia.) in slabs was achieved both during the steady state and the transient state of casting. Melt splash accretions were also decreased to one third. This is because chilling of the melt near the meniscus at the inner wall of the pouring tube was decreased, due to the melt turbulence caused by submerged Ar injection. In addition, injected Ar above the meniscus prevented melt splash from attaching to the inner wall.

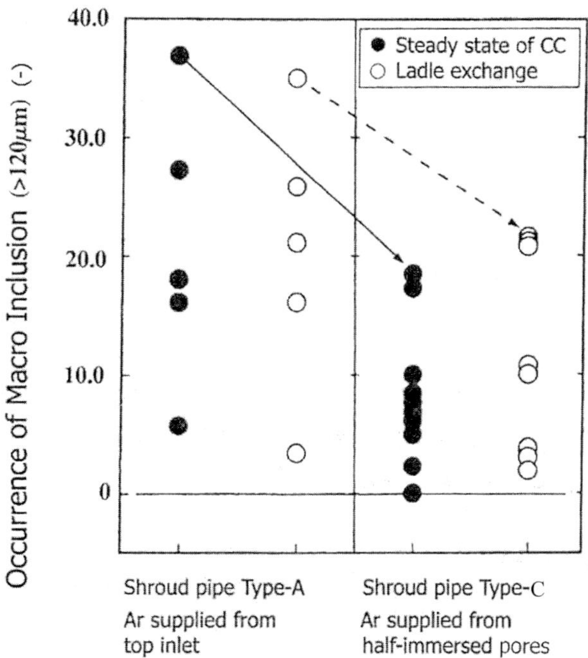

Figure 8.22: Reduction in macro inclusions due to improvement in shrouding pipe and Ar injection system. [Ref. 15]

8.9 Concluding Remarks

The tundish, over the years, has substantially improved to deliver clean steel melt without contamination at a high rate and steady temperature to mold for the steady state casting under a wide range of operating conditions. For temperature control, active measures like tundish heaters and a tundish hot cycle have solved problems arising during the transient periods, particularly for low superheat casting. Also, refractory consumption of the tundish has decreased drastically by single-tundish hot cycling with preheating under a reducing atmosphere.

However, the occurrence of exogenous macro inclusions in transient periods has not yet been eliminated for difficult steel grades at a high throughput rate. Reoxidation by air and emulsification of ladle slag in the inlet compartment of the tundish has been difficult to avoid under varying conditions of tundish operation, even with the use of impact pads of various configurations. Some of the measures described in this chapter have already been implemented in industrial operation, but others are in their trial stages, and require further improvements. With regard to process problems arising from the macro inclusions, clogging of the SEN has long been the most serious problem to be resolved. As the cause and mechanism for the occurrence of nozzle clogging have been understood, tundish-related countermeasures to prevent the clogging are also being worked out.

All of the problems mentioned in this chapter are by their nature moving targets. As the requirements for the product quality gets more demanding, more effective countermeasures should be developed and implemented. Optimization of quality, productivity, and cost needs to be carefully addressed in developing any countermeasures. Operating conditions for each caster are different, and hence there seems to be no universal design of a tundish for all casters.

General trends show that a large volume tundish, consisting of inlet and outlet compartments with reasonable connection in between and sufficient melt depth in the outlet compartment, can deliver clean melt to the mold without appreciable contamination during steady state casting. A long nozzle or shrouding pipe should be used for the inlet compartment, and the outlet compartment should have inert atmosphere or covered with a tundish flux of suitable composition. Sophisticated furniture (flow control devices) can be removed when air reoxidation and slag emulsification are suppressed in the inlet compartment for cost reasons and for convenience in hot cycling.

For the transient periods, mandatory practices include (1) removal of filler sand before ladle opening; (2) flushing of inlet- and outlet-compartment with Ar gas before and during the ladle opening of every heat in a sequence; (3) use of long nozzle or shrouding pipe; (4) a filled start after reasonable elapsed time; and (5) prevention of vortexing ladle slag during melt draining. These issues have been addressed in the last three decades, yet countermeasures are incomplete due to operational difficulties. However, an industrial tundish operation for bearing steel with a complete inertization of the tundish, combined with leaving some melt in ladle to avoid slag vortexing, has made macro inclusions of reoxidation and emulsification origin virtually nonexistent.

This clearly indicates that the mechanisms and countermeasures to prevent harmful macro inclusions are correct, but that the execution of the countermeasures has been incomplete in practice for conventional casters. Tundish design and operation are always determined by a compromise of cost with quality and productivity. The market demands improvements towards better quality without increasing cost, and tundish technology can develop to fulfill these demands.

References

1. H. Kimura, A. Uehara, M. Mori, H. Tanaka, R. Miura, T. Shirai, and K. Sugawara, *Shin-Nittetsu Giho (Nippon Steel Corp. Techn. Report, in Japanese),* 1994, 351, 21-26.
2. K. Takase, K. Misawa, K. Amada, M. Amano, N. Konno, A. Uehara, and Y. Yamamura, *CAMP-ISIJ,* 1997, 10, 138.
3. Y. Miki, H. Shibata, N. Bessho, Y. Kishimoto, K. Sorimachi, and A. Hirota, *Tetsu-to-Hagane,* 2000, 86, 239-246.
4. H. Yin, H. Shibata, T. Emi, and M. Suzuki, *ISIJ International,* 1997, 37, 936-945.
5. K. Okumura, M. Ban, M. Hirasawa, M. Sano, and K. Mori, *Tetsu-to-Hagane,* 1994, 80, 201-206.
6. S. Takeuchi, S. Idogawa, K. Sorimachi, and T. Sakuraya, *CAMP-ISIJ,* 1992, 5, 206.
7. A. Idogawa, S. Takeuchi and N. Bessho, *CAMP-ISIJ,* 1995, 8, 211.

8. M. Morishita, T. Tai, K. Ayata and J. Katsuta, *CAMP-ISIJ*, 1997, 10, 832.

9. S. Nara, K. Sorimachi, N. Bessho, K. Kariya, and R. Asaho, *CAMP-ISIJ*, 1997, 10, 764-767.

10. S. Kittaka, S. Wakida, T. Sato and M. Miyashita, *Shin-Nittetsu Giho (Nippon Steel Corp. Techn. Report, in Japanese)*, 2005, 382, 16-20.

11. S. Yamada, T. Shiragami, N. Kitada and N. Ono, *CAMP-ISIJ*, 2005, 18, 214.

12. K. Hara, T. Nakagawa, S. Yuhara, H. Osanai, T. Abe and Y. Wakatsuki, *CAMP-ISIJ*, 2000, 13, 108.

13. T. Abe, K. Hara, T. Nakagawa, K. Kadota, T. Fujimura Nd H. Nomura, *CAMP-ISIJ*, 2000, 13, 109.

14. S. Kooriyama, H. Tsurumaru, T. Hori, M. Mikuni, H. Uehara and A. Shiroyama, *CAMP-ISIJ*, 2005, 18, 216.

15. N. Kasai, H. Yamazoe and M. Iguchi, *Tetsu-to-Hagane*, 2005, 91, 763-768.

Subject Index